GLOBAL ENVIRONMENTAL CHANGE AND LAND USE

Global Environmental Change and Land Use

Edited by

A.J. Dolman

Free University Amsterdam,
The Netherlands

A. Verhagen

Plant Research International,
Wageningen University and Research Centre, The Netherlands

and

C.A. Rovers

National Research Programme on Global Air Pollution and Climate Change,
Bilthoven, The Netherlands

KLUWER ACADEMIC PUBLISHERS
DORDRECHT / BOSTON / LONDON

A C.I.P. Catalogue record for this book is available from the Library of Congress.

ISBN 978-90-481-6308-3

Published by Kluwer Academic Publishers,
P.O. Box 17, 3300 AA Dordrecht, The Netherlands.

Sold and distributed in North, Central and South America
by Kluwer Academic Publishers,
101 Philip Drive, Norwell, MA 02061, U.S.A.

In all other countries, sold and distributed
by Kluwer Academic Publishers,
P.O. Box 322, 3300 AH Dordrecht, The Netherlands.

Printed on acid-free paper

Contents

7

Climate change and food security in the drylands of West Africa 167
A. Verhagen, A.J. Dietz, R. Ruben, H. van Dijk, A. de Jong, F. Zaal, M. de Bruijn and H. van Keulen

8

Land-use changes induced by increased use of renewable energy sources 187
S. Nonhebel

Preface

The interaction between environmental change and human activities is complex, requiring the concepts and tools of a number of disciplines for its effective analysis. Land-use and land-cover change has only recently become a topic susceptible to scientific research, as these concepts and tools have been developed and made available. Rooted in a broad community concerned with global change, systematic research has begun into land-use systems at different scales and interactions, and their links with global cycles of water, nitrogen and carbon are being explored. Partly based on research initiated by the Dutch National Research Programme on Global Air Pollution and Climate Change (NRP), this book touches upon various land-use and land-cover issues in relation to global environmental change. In addition to the biogeochemical cycles, land as a carrier for functions of economic activities, food and fibre production and energy production via biomass are discussed. Crucial in studying land use is human behaviour and man-environment interaction at different scales.

Land-use and land-cover change is an important contributor of greenhouse-gasses as these activities directly interfere with the carbon, nitrogen and water cycles. These cycles are connected through numerous feedback loops. The interface of land-use and climate is essentially determined by the interaction of man and the environment. Man uses land primarily to produce food; a relatively small area is needed for urban development. All cycles, and in particular, the nitrogen cycle, are at various scales determined by increased food production or the increase of agricultural land. The disturbing effects on climate and climate variability are obvious. Responses to this knowledge prompted policy makers, industry and science to define strategies to reduce emissions and adapt to climate change. Poorly designed strategies, be they at the scale of the individual, landscape or nation, may put stress on systems and increase disparity between groups and countries. Coordinated management of system components at all scales is needed to address key issues related to global environmental change.

Combining the various parts, as described in this book, is still in its infancy but scientific efforts to understand the interactions among components of the biogeochemical cycle, notably nitrogen and carbon, and interactions with the climate system, are needed to identify vulnerable components and systems.

Defining balanced land use strategies requires knowledge of the physical components, but understanding of economic drivers and behavioral aspects of the decision-makers are essential as well. The communities of physical and social scientists have only just begun to combine forces to address the most pressing issues facing the world: sustainable development.

The current book highlights a number of key issues dealing with land use in a changing world. It attempts by no means to cover all aspects, rather it focuses on a selected number of aspects drawing from research projects in the NRP, placing these in a broader, global, perspective. By doing this, a state of the art review of the human impact on changes of land use is obtained. The consequences on climate and hydrology are explored in separate chapters. The resilience of man and ecosystems to adapt to a changing climate or mitigate some of the worst effects is explored in two chapters on the use of biomass energy and food.

HAN DOLMAN, JAN VERHAGEN AND ILSE ROVERS
AMSTERDAM/WAGENINGEN/BILTHOVEN, APRIL 2002

Acknowledgements

This book comprises the efforts of a considerable number of people who were partly funded by the Dutch National Research Programme on Global Air Pollution and Climate Change (NRP). In the first place we would like to thank the authors who found the time and courage to write the individual chapters. Each chapter was reviewed by at least two reviewers. Our warm thanks go to those anonymous reviewers.

The Program Office provided support and reminded us time after time that the book needed to be finished. To Bert Jan Heij, Wilko Verweij, Marcel Kok and Ottelien van Steenis, many thanks.

I

INTRODUCTION

Chapter 1

LAND USE AND GLOBAL ENVIRONMENTAL CHANGE

A.J. Dolman
Vrije Universiteit Amsterdam

A. Verhagen
Plant Research International, Wageningen University and Research Centre

1. Global environmental change

Changes in land use and land cover have contributed substantially to the increased CO_2 content of the atmosphere, have exacerbated shortages of water, have substantially changed biogeochemical cycles on earth, and are causing dramatic losses of biodiversity around the globe. The combined effects of these forces on global climate, biodiversity, water availability and ecosystem-inxxecosystem vulnerability amongst others are generally denoted as global environmental change. Land use links human activities to land cover. Changes in land use and land cover are largely related to changes in the exploitation of land. The demand for agricultural land has increased, mainly triggered by the growing and changing food demand from an ever-increasing population. In areas where land is scarce the need to maintain food production with growing demand is achieved via technological changes in land use accomplishing higher returns per the area of land. Where land was abundant, land conversion remained the main strategy. The growing demand on scarce land resources has resulted in additional environmental stress. On the other hand, the intensification of land use, which was successful in increasing food production, has been damaging to the environment. Today's environmental challenges arise both from the lack of development and from the unintended consequences of some forms of economic growth (Brundtland *et al.*, 1987). Figure 1.1 (Vitousek *et al.*, 1997) shows examples of the impact of man's activities in the form of estimates of the percentage transformation of the earth's surface. Land-use and land-cover changes are thus prime and central phenomena in shaping today's world and its climate.

3

A.J. Dolman et al. (eds.), Global Environmental Change and Land Use, 3-13.
© 2003 *Kluwer Academic Publishers.*

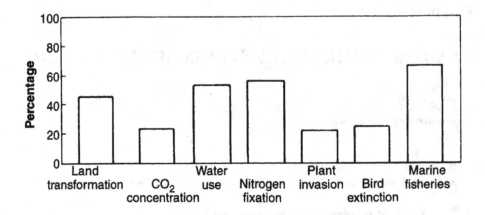

Figure 1.1. Human alteration of several components of the earth system expressed as percentage of the original state or total process. About 45% of the available land is transformed by human use, 50% of the earth's accessible fresh water resources are used and 20% of the current CO_2 concentration of the atmosphere is a direct result of human enterprise. 60% of the terrestrial N-fixation is human caused and 68% of the marine fish resource is depleted, overexploited or near the limit of exploitation. 20% of the plant species of Canada is introduced from elsewhere and more than 20% of the bird species on earth has become extinct in the last two millennia as a consequence of human activity, after Vitousek *et al.*, (1997).

Man, however, is both agent and victim of global environmental change. Human activities, primarily the combustion of fossil fuels, have increased the concentrations of greenhouse gasses in the atmosphere. The effects of the corresponding changes in climate and climate variability are expected to be greatest in the developing countries (IPCC WGII, 2001). A key question therefore is, if man can also act as a mitigator of global change and control aspects of for instance climate change. The Kyoto protocol and several other international environmental agreements arising from the Earth Summit of Rio de Janeiro in 1992 at least raise the hopes that there may be political support for this in the near future. This is not to say that it will be technologically achievable, but it is obvious that good land management is a prerequisite for a transition towards a more sustainable use of the earth's resources.

Understanding the socio-economic causes and main driving forces of land-use and land-cover changes and the impacts of these changes on the major biogeochemical cycles (water, nitrogen, carbon) and climate are crucial in identifying successful adaptation and mitigation strategies. This book aims to provide an overview of a number of important aspects of land use and land cover change. It purposely does so from the broad perspective of global change and thus attempts not to isolate single effects. The book does not aim to present an integrated assessment of social, economic, cultural causes of land-use change and the impacts on climate. What it does try to do is to highlight a number of important aspects of land use and land-use change. We achieve this by a three-step approach where the first two chapters describe the general concepts of land-use and land-cover change, including the socio-economic drivers. The second block deals with the relation between climate and land-use and land-cover change and the emission of major greenhouse gasses. The third block is about impact, adaptation and mitigation. The impact of global environmental change on global water resources and water availability for different types of land-use is discussed. An exploration of impacts on food security follows, presenting the case of Sub-Saharan West Africa addressing both biophysical impacts and societal responses to changing environmental conditions. It closes with an analysis of possibilities of the use of biomass production to substitute for fossil fuels. All these forms of land use have claims, often competing, on the available scarce land.

2. Understanding the driving forces of land use and land-use change

Global environmental change and climate change can only be understood when the major causes of land-use change are understood. These are often culturally, economically and sociologically shaped. In Europe considerable amounts of land have been cleared of their forests in the past two millennia, and virtually all land today is man managed, in more or less intensive ways. Land-use change in developed regions therefore deals mainly with changes in production systems (crops, fertiliser and pesticide, number of animals) and less with considerable changes in land-cover, such as a shift from agriculture to forestry. These tend to be slowly changing. For instance, analysis of the main causes of land-use change in Europe suggests a strong influence of the Common Agricultural Policy of the EU and a small relative decline in agricultural land over the last 30 years. This highlights an important aspect of land-cover change: it is often, if not always, economically motivated. Environmental concerns hardly play a role. This slowly changing nature of land-use and land-cover change in the developed world is in sharp contrast with areas in the developing world.

In the tropics, demand for agricultural land continues to be one of the main driving forces for land-cover changes such as deforestation of the tropical rainforests, and cultivation of marginal lands. According to the latest 1999 FAO assessment of the world's forest resource, the total area of forests decreased in 1990–1995 by 56.3 million hectares. This number hides the difference between developed and underdeveloped areas: in the underdeveloped areas 61.5 million hectares were lost, and this was partly compensated by a small increase of 8.8

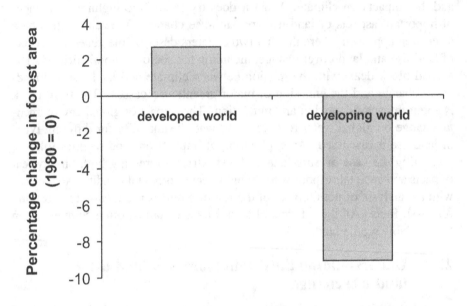

Figure 1.2.   Global deforestation between 1980 and 1995. (source: WRI, 2000, Pilot Analysis of Global Ecosystems: Forest Ecosystems).

million hectares in the developed world (Figure 1.2). The main causes of the loss of forests are conversion of land to agriculture and the development of large infrastructural facilities in developing countries. In the developed world, forests are allowed to grow on agricultural land that was taken out of production.

Predicting the future trend of land use, particularly in developing countries, requires studying the interaction of micro- and macro-economic processes with biophysical properties of soil and climate. Figure 2.1 (Chapter 2) shows the linkage of these driving factors. An important set of issues becomes clear when looking at this myriad of relations: scale, organisation level, hierarchy and analysis. Several processes occur at different scales and this may not always correspond to the scale at which social or legal institutions are organised. Furthermore, the level of analysis applied, or the level at which data is available, is not always the correct one for answering the questions posed, which may originate from a different level of scale. Consequently, new methodologies, such as multi scale analysis of land-use change have been formulated, tested and applied in the field (Veldkamp and Fresco, 1996).

In the real world, land-use change, be it applying new fertilizer or conversion of forest to agricultural land, is executed by individual people or farmers, who make their choices based on a number of rational and irrational considerations.

Economics and sociology aid by trying to understand these choices, and building on that analysis, help to identify new incentives that may eventually become part of public policy. They aim to analyse where activities are allocated and which plots can be allocated to a particular form of land-use. It is here that biophysics meets economics. The suitability of the land is foremost an issue of soil quality, climate conditions and available infrastructure. The question as to where activities may be located is one of regional economy. It is a question of spatial dimension in economic activity and deals with issues such as distance to market, transportation costs, tax incentives etc. To properly understand the economics of land use and land-use change one needs to study both.

An important new area of economic research is the development of economic incentives to mitigate climate change by reducing emissions or storing greenhouse gasses. Examples are the introduction and acceptance of new technologies for staple crops such as rice that may reduce methane emissions from rice paddies, and the use of plantation forestry or agroforestry to sequester CO_2. The existing low cost estimates of the latter are a prime motivation for developed countries to seek considerable amounts of the assigned Kyoto reductions abroad (Chapter 3). Associated problems such as leakage (the option that avoiding deforestation in one area may lead to deforestation in an adjacent area) require above all economic incentives for the people to earn a living, other than that from the forest. Mitigating the CO_2 problem in a world with a global economy requires also analysis of the global context of emissions, its economic drivers (Lubbers *et al.*, 1999).

3.   Changing the global biogeochemical cycles

Globally the land surface is a non-negligible factor in the carbon and nitrogen cycles. As shown in Figure 1.1, 20% of the current atmospheric CO_2 concentration can be attributed to human influence and the amount of nitrogen fixated has changed by almost 60% due to human action. Land use thus plays an important part in the biogeochemical cycles. At the earth's surface vegetation and soil interact with the atmosphere to produce complex flows and exchanges of water, nutrients and heat. Carbon and nitrogen change from fast rotating pools and atmospheric storage to more slowly rotating pools and may eventually get locked up in long term, very slowly changing pools.

Table 1.1 shows the impact of three major greenhouse gasses on the radiative forcing of the atmosphere. All three gasses are formed or taken up at the earth's land surface. The table also shows the increase for these gasses since the industrial revolution. The increase in all three gasses is considerable, but as suggested by Hansen *et al.* (2000) in the short term, a considerable amount of reduction in global warming may be obtained if the emissions of CH_4 and N_2O are curbed. In the longer term, CO_2 levels have of course to be reduced by introducing new energy carriers and improving energy efficiency. Emissions of CH_4 and N_2O are mainly caused by agricultural practises. Agricultural practises increase the amount of CH_4 and N_2O, but also of CO_2, mainly because humans manipulate the carbon and nitrogen cycle to produce food.

Table 1.1. Increase in the three main greenhouse gasses, their global warming potential (compared to CO_2) and contribution to radiative forcing (source: IPCC Third Assessment Report).

	Period		Global Warming	Increase in radiative
	1000–1750	*present*	Potential	forcing (Wm^{-2})
CO_2	280±6 (ppm)	368 (ppm)	1	1.5
CH_4	750±60 (ppb)	1750 (ppb)	21	0.5
N_2O	270±10 (ppb)	316 (ppb)	310	0.15

CO_2 remains the most important greenhouse gas. The political importance of the land surface has increased dramatically since the Kyoto protocol recognises the potential of the land to sequester carbon. At the Third Conference of the Parties (COP-III) in Kyoto, for the first time, the parties of the industrialised world (ANNEX-I) agreed to limit their emissions of greenhouse gasses by a fixed amount. To achieve this reduction the protocol included the possibility of using the terrestrial biosphere to sequester carbon (in paragraphs 3.3. and 3.4 of the Kyoto protocol). The exact rules and modalities under which 'sinks' will be included was to be decided at the Sixth Conference of the Parties (COP-VI) in the Hague, but the parties failed to achieve agreement.

The Intergovernmental Panel on Climate Change (IPCC, Third Assessment Report) estimates that of the atmospheric increase in CO_2 since 1860 (the start of the industrial revolution), roughly one-third is caused by land use and land-use change. In the global carbon balance (Figure 1.3) the source from land-use change and the sink from terrestrial uptake roughly balance (1.6 vs 2.3 Gton in the 1990s, 1.7 and 1.9 Gton in the 1980s). This leaves a small net residual uptake of 0.2 and 0.7 Gton respectively in the 1980s and 1990s.

Terrestrial biospheric sinks are dramatically different from other terms in the carbon balance. Emissions from fossil fuel and cement production show a steady, almost predictable slow and steady increase. Terrestrial biospheric uptake in contrast depends on weather, the amount of precipitation, temperature and radiation, and shows variability associated with the main weather patterns. On top of this pulse events, like forest fires, major storms, and insect plagues are continuously changing the carbon balance of the biosphere. This leads to considerable variations in the terrestrial uptake at the scale of a few years to a decade. Land management and changes in land-cover can however also be

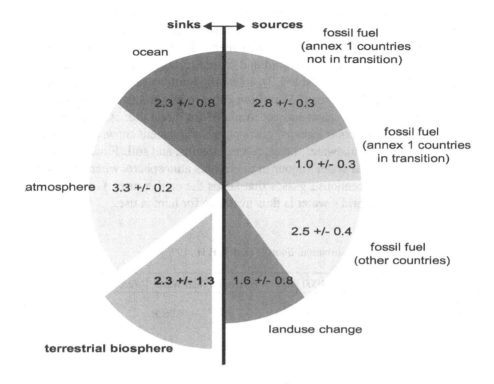

Figure 1.3. The annual global carbon balance in the 1990s. Fossil fuel emissions are 6.3 Gton C, Storage in the atmosphere 3.3 Gton C, Ocean uptake 2.3 Gton C, Land use change (primarily deforestation) 1.6 Gton C, terrestrial uptake 2.3 Gton C. The net terrestrial uptake (uptake minus deforestation) is 0.7 Gton C. Source: IPCC, Special report on Land Use, Land Use Change and Forestry, 2000.

used to mitigate the effects of climate change. The major way in which this can be achieved relates to stopping or slowing down current deforestation. Other ways to absorb more CO_2 include the planting of new forests and changing management in agricultureagriculture and forestry. The cost effectiveness of these measures appears generally to be below the reduction costs for other more technological measures, but there are doubts on the sustainability of sinks. Chapter 5 discusses these issues and provides examples of potential for CO_2 reduction through the use of sinks.

4.   Changing water availability and climate land use interactions

The driving forces of land use continue to change the major forms of land use. The impact of land-use and land-cover change on the atmosphere and availability of water resources is however considerable. This requires us not only to look at the economics and sociology of land-use change, but also at the impact changes may have on our natural environment. In this environment water is a key resource. Water, in both liquid and frozen forms, covers approximately 75% of the Earth's surface (Table 1.2). In all, the Earth's water content is about 1.39 billion cubic kilometres (331 million cubic miles) and the vast bulk of it, about 96.5%, is in the oceans and not available for direct use. Approximately 1.7% is stored in the polar icecaps, glaciers, and permanent snow, and another 1.7% is stored in groundwater, lakes, rivers, streams, and soil. Finally, a thousandth of 1% exists as water vapour in the Earth's atmosphere, where it acts as one of the major greenhouse gasses that keeps the earth warm. Only a small percentage of the world's water is thus available for human use.

Table 1.2.   Global water distribution. Source: Gleick, P. H., 1996.

	Volume (1000 km^3)	Percent of Total Water	Percent of Fresh Water
Oceans, Seas, & Bays	1,338,000	96.5	–
Ice caps, Glaciers, & Permanent Snow	24,064	1.74	68.7
Groundwater	23,400	1.7	–
Fresh	(10,530)	(0.76)	30.1
Saline	(12,870)	(0.94)	–
Soil Moisture	16.5	0.001	0.05
Ground Ice & Permafrost	300	0.022	0.86
Lakes	176.4	0.013	–
Fresh	(91.0)	(0.007)	0.26
Saline	(85.4)	(0.006)	–
Atmosphere	12.9	0.001	0.04
Swamp Water	11.47	0.0008	0.03
Rivers	2.12	0.0002	0.006
Biological Water	1.12	0.0001	0.003
Total	1,385,984	100.0	100.0

Water is arguably the single most important resource on earth; 50% of the earth's fresh water reserves are impacted by human actions. Shortages may turn into drought and excess rainfall may cause damaging floods. Amongst the highest research priorities in Earth science are the potential changes in the Earth's water cycle due to climate change, if only because water is by far the most important greenhouse gas.

The water cycle is a complex cycle where water from rains is transported through runoff into rivers and ends up in oceans and lakes, where it is evaporated into the atmosphere and eventually produces rain again. The land surface provides a critical role in this cycle as it is the level at which precipitation is redistributed into evaporation, runoff or soil moisture storage. The amount of evaporation influences the land surface energy balance (the surface temperature of the land) that impacts the temperature and moisture content of the atmosphere above the land. These interactions provide critical feedbacks to the water cycle and are the prime cause of the conjecture that deforestation, or degradation of previously green land, may have a decreasing influence on rainfall. At the global scale, climate model simulations show that land-cover change, in particular the conversion of forest to non-forest or agricultural land, can have a considerable effect on the climate. In the Amazon, deforestation can potentially reduce rainfall below levels, where regrowth of forest may become impossible. For the Sahel, studies with coupled climate-ecosystem models provide new understanding of the causes of rainfall variability in that region. The interaction of the land-cover with the atmosphere is the subtle mechanism whereby nature controls ecosystem diversity and climate. The precise mechanisms, through which this feedback occurs, are the subject of Chapter 5.

There is however also a direct impact of the land surface on the availability and distribution of water. This is the subject of Chapter 6. It has been known for some time now that different types of land-cover have different requirements for water. This is also reflected by the distribution of the global vegetation cover: arid and semi-arid regions have less annual precipitation than for instance tropical rainforests. However, water availability for groundwater is influenced by the simple water balance equation, where the residual water for groundwater replenishment is obtained by subtracting soil evaporation and transpiration and runoff from the precipitation. Generally tall vegetation, such as forests, use more water, because they intercept rainfall, which is subsequently evaporated back into the atmosphere without reaching the soil surface. The amounts vary: from 9% to 45% for tropical rainforest and from 10% to 30% for temperate forests. The higher aerodynamic roughness of tall vegetation facilitates this rapid evaporation. The effects of deforestation are to first order equal to the difference in interception loss between forest and transpiration. Deforestation, particularly on sloping ground, is believed to contribute to increased runoff and increased erosion in tropical and semi-arid regions. Higher water use through forest is also possible because roots reach deeper and may access more soil moisture. However, changing crops also induces changing water requirements and locally there are effects on groundwater replenishment and runoff for crops other than forests.

5.   Agriculture and options for mitigating climate change

Agricultural demand for water puts the largest claim on the world freshwater resource. Associated with the growing world population, not only water for food production will increase, but also direct consumption and changes in consumption patterns will affect the demand for water. In industrialised and transition economies the claim from urban centres and industries will become more important (Chapter 6). Already large numbers of people live in water scarce areas, Areas with low and erratic rainfall will be more easily affected by climate change. With increasing demand on natural resources and food, climate change may add further pressure on agricultural systems, particularly in the developing world. This puts great emphasis on the resilience, adaptability and sustainability of these land use systems within a changing climate. Chapter 7 deals with the impacts of rainfall variability in Sub Saharan West Africa on agricultural production and identifies coping strategies, combining both biophysical and socio-economic aspects.

The use of biomass to substitute for fossil fuel is another important usage of the biosphere in mitigating the effects of climate change. Ultimately this is the only sustainable use of the biosphere, as there are reasons to doubt the long-term capacity of the existing biosphere to continue absorbing CO_2. Scenario studies with integrated assessment models, however, show that to substitute all oil by biofuels would require 50% more agricultural land globally than currently in use for all other agricultural purposes (Alcamo, *et al.*, 1994). For Europe it was estimated that 17% of the European land area would be needed to provide Europe completely with biomass fuels. It is unlikely that this is a realistic option. It does however show an important aspect of land use. To substitute biofuel for oil requires land and this opens up immediately competition with demands for other functions, in particular agriculture (Chapter 8). The demand for space this creates may however be in conflict with demands to grow more food. With increasing demand on natural resources and food, climate change may add further pressure on agriculturalagriculture systems, particularly in the developing world. This puts great emphasis on the resilience, adaptability and sustainability of these land-use systems within a changing climate.

References

Alcamo, J. (ed.) (1994) IMAGE 2.0: Integrated Modeling of Global Climate Change. Dordrecht, The Netherlands: Kluwer Academic Publishers.

Brundtland *et al.* (1987) Our Commom Future. The world commission on environment and development. Oxford University Press.

Gleick, P. H., 1996: Water resources. In Encyclopedia of Climate and Weather, ed. by S. H. Schneider, Oxford University Press, New York, vol. 2, 817–823.

Hansen, J., Sato., M., Ruedy, R, Lacis, A., Oinas, V. (2000) Global warming in the twenty-first century: An alternative scenario. Proc. Nat. Ac. Sci. USA 97, 9875–9880.

IPCC WGII (2001) Summary for Policymakers. Climate Change 2001: Impacts, adaptation, and vulnerability. http:www.ipcc.ch.

Lubbers, R., de Zeeuw, A., Koorevaar, J., de Nooy, M. (1999) CO_2 als extern effect van de globaliserende economie- een wereldwijd gedifferentieerd probleem. In: Kok, M and de Groot, W. (Eds). Een klimaat voor verandering, tien essays over klimaat en klimaatbeleid. NRP, Bilthoven, 71–90.

Veldkamp A. and Fresco L.O. (1996) CLUE-CR: an integrated multi-scale model to simulate land use change scenarios in Costa Rica. Ecological Modelling 91, 231–248.

Vitousek, P.M. Mooney, H.A., Lubchenko, J. and Mellilo, J.M. (1997) Human domination of Earth's ecosystems. Science 277, 494–499.

WRI (2000) Pilot Analysis of Global Ecosystems: Forest Ecosystems.

II

LAND-USE AND LAND-COVER CHANGE: GENERAL CONCEPTS

Chapter 2

METHODOLOGY FOR MULTI-SCALE LAND-USE CHANGE MODELLING: CONCEPTS AND CHALLENGES

P.H. Verburg
Department of Environmental Sciences, Wageningen University and Research Centre

W.T. de Groot
Centre of Environmental Science, Leiden University

A.J. Veldkamp
Department of Environmental Sciences, Wageningen University and Research Centre

1. Introduction

Land cover is defined as the layer of soils and biomass, including natural vegetation, crops and human structures that cover the land surface. Land use refers to the purposes for which humans exploit the land-cover (Fresco, 1994). Land-cover change is the complete replacement of one cover type by another, while land-use changes also include the modification of land-cover types, e.g., intensification of agricultural use, without changing its overall classification (Turner II *et al.*, 1993).

The individual human activities that lead to land-use changes meet locally defined needs and goals, but aggregated they have an impact on the regional and global environment (Turner II, 1994; Ojima *et al.*, 1994). They may affect biodiversity, water and radiation budgets, trace gas emissions and other processes that, cumulatively, affect global climate and biosphere (Intergovernmental Panel on Climate Change, 1997).

Land-use change is directly linked to the theme of 'transition to a sustainable world'. With origins in the Brundtland report and increasingly embedded in global change science agendas, the overarching concern is achieving sustainability in a warmer, more crowded, and more resource-demanding world characterised by unexpected and extreme events. This transition requires an improved understanding of the trajectories of land-use change that entail posi-

A.J. Dolman et al. (eds.), Global Environmental Change and Land Use, 17-51.

tive or negative human-environment relationships (Turner II and Brush, 1987; Holling and Sanderson, 1996).

A better understanding of land-use change is essential to assess and predict its effects on ecosystems and society. Greenhouse gas emission inventories and reduction objectives need to include changes in emissions caused by land-use change (Verburg and Denier van der Gon, 2001). Rates and patterns of land-use change need to be understood to design appropriate biodiversity management. Areas of rapid land-use change need to be identified to focus land-use planning on the appropriate regions.

The objective of this chapter is to provide an overview of research methodologies for understanding and modelling land-use changes. Based on this overview and the science plan of the international Land-Use and Land-Cover project (LUCC; Lambin *et al.*, 2000a), we will discuss and introduce future directions of land-use change research. In the next paragraph a short review of land-use change processes and their causes, driving factors, at different scales is provided. Different research approaches to identify and quantify the driving factors of land-use change are described. Application of the thus obtained understandings in dynamic models is also possible in many different forms. Three different modelling approaches are described and discussed. Overall, we will emphasise the problems and solutions of multi-scale modelling, i.e., modelling that takes into account that the causes of land-use change are not simply the aggregates of the causes of land-use change in a number of localities only. Examples of two methodologies for land-use change research, developed within the Dutch National Global Change Research Programme (NRP) and originating from different disciplines, are given as illustration. The chapter ends with a discussion of possible applications of the information obtained from land-use change models and addresses how future research on land-use change can make better use of the richness of methods and insights available at the interface of the different disciplines involved.

2.   Multi-scale driving factors of land-use change

Land-use and land-cover change are the result of many interacting processes. Each of these processes operates over a range of scales in space and time. With the term scale we refer to the spatial, temporal, quantitative, or analytic dimensions used by scientists to measure and study objects and processes. All scales have extent and resolution. Extent refers to the magnitude of a dimension used in measuring (e.g., area covered on a map), whereas resolution refers to the precision used in this measurement (e.g., grain size). For each process important to land-use and land-cover change, a range of scales may be defined over which it has a significant influence on the land-use pattern (Meentemeyer, 1989; Dovers, 1995). These processes are driven by one or more of these variables that influence the actions of the agents involved in land-use and land-cover change. Often, a distinction is made between social and biophysical driving forces. Figure 2.1 gives a summary of driving forces and the range of scales where the most prominent influence of the individual factors on land use is commonly identified. Reviews of the driving factors of land-use change are given by Turner II *et al.* (1993) and Lambin *et al.* (2000b). Often, the range of spatial scales over which the driving factors and associated land-use change processes act, correspond with levels of organisation in a hierarchically organised system characterised by their rank ordering in the hierarchical system. Examples of levels include organism or individual, ecosystem, landscape and national or global political institutions. Many interactions and feedbacks exist among processes at different levels of organisation. Hierarchy theory suggests that processes at a certain scale are constrained by the environmental conditions at levels immediately above and below the reference level, thus producing a constraint 'envelope' in which the process or phenomenon must remain (O'Neill *et al.*, 1989). Land-use change research would become dramatically complicated if all hierarchical relations among all driving factorsdriving factors would have to be incorporated.

How should we study this complexity? The above described complexity of interacting processes of land-use change suggests that finding appropriate methodologies for studying land-use and land-cover change is not an easy task (Wilbanks and Kates, 1999). The first important consideration is that scales of analysis usually do not correspond to levels of organisation (O'Neill and King, 1998). A sociological survey of political opinions, for instance, refers to the organisational level of individual people, but has, at the same time, usually the spatial extent of a nation. A study of common property management in a village, on the other hand, is at a group level of organisation but at a local spatial scale. And finally, an organisation such as the World Bank may be studied as a single social entity (actor), i.e., at the micro level of organisation, and may be viewed as responding to and influencing factors at the global scale. This enables us to better analyse, for instance, how phenomena that are 'micro' in terms of organisational level may have consequences on a spatially regional scale. Or the reverse, that a phenomenon such as the price of fertilizer, which is a 'macro' phenomenon in terms of social organisation, because it is an emergent property of aggregated supply and demand, exerts an important influence at the spatial scale of the field of an individual farmer.

In our research we usually opt for one level of analysis exclusively, without considering the range of other alternatives. Often, this choice is based on arbitrary, subjective reasoning and not reported explicitly (Gibson *et al.*, 2000; Watson, 1978). In some studies, the choice for the scale of observation is based on the assumption of discrete levels of organisation, e.g., communities

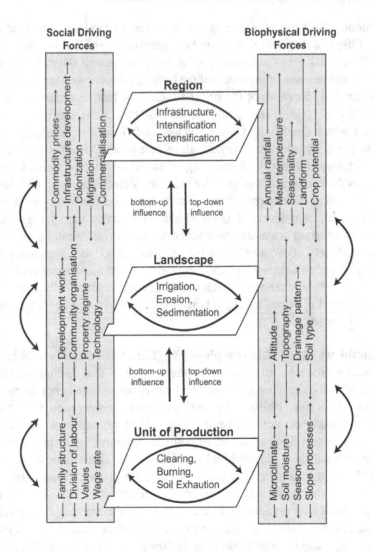

Figure 2.1. Common driving factors and range of scales where these driving factors have the most prominent influence on land-use change. Adapted from Turner II *et al.*, 1995.

or ecosystem patches. However, as a result of the many interacting processes, each at different levels of organisation, ecosystems and land-use systems rarely or never are restricted to a single scale that can be regarded as correct or optimal for measurement and prediction (Gardner, 1998; Geoghegan *et al.*, 1998; Allen and Star, 1982; Levin, 1992). Although for a specific data set optimal levels of analysis might exist where predictability is highest (Veldkamp and Fresco, 1997; Goodwin and Fahrig, 1998), unfortunately these levels are not consistent through analysis. Therefore, it might be better not to use a priori levels of observation, but rather extract the observation levels from a careful analysis of the data (O'Neill and King, 1998; Gardner, 1998). Also with respect to the choice of variables selected for analysis one needs to be cautious. It is often assumed that more parsimonious explanations exist when proposed causal factors operate at the same spatial scale as the observed land-use changes. Turner (1999) calls this scaling parsimony. Often, an accepted reason for excluding a locally important factor from a regional analysis is that the local variations caused by the factor are distributed such that its aggregate effect on regional land use is small. Due to data limitation problems, rigorous application of this averaging rationale is rare. Therefore, many social factors are viewed a priori as 'locally specific' and excluded from analysis.

The problem is, however, that the method used to choose 'regionally relevant' variables is rarely described, and local studies conducted within a region are rarely referenced. The regional or local slices through the causal web over multiple scales are derived not from an inherent spatial-scaling law of society, but from an analyst's choice, which is socially constituted. Based on his findings in the Sahelian region, Turner (1999) argues against a priori categorisation of certain types of social change as 'local', leading to their exclusion, by scaling parsimony, from consideration as causal agents in regional analyses of land-use change. In his study he shows that gender relations in rural Africa, often labelled as 'locally specific' have important regional consequences for changes in the composition of the livestock population, while changes in price or livestock productivity, often classified as regionally important variables, have limited effect. This brings us to the insight that single-scale approaches do not suffice to obtain a good understanding of land-use change.

Researchers in the field of land-use studies share the formidable task of coming to grips with the complex causal web linking social and biophysical processes (Turner II, 1997). For the analysis of multi-scale dynamics of complex systems new methodologies need to be developed. The next section describes research approaches for land-use change analysis as developed by different disciplines and discusses how multiple scales/levels can be incorporated within these approaches.

3. Research approaches in land-use studies

Different research approaches, strongly divided by scientific discipline and tradition, have emerged in the field of human-environment interactions. Researchers in the social sciences have a long tradition of studying individual behaviour at the human-environment interface at the micro scale, some of them using qualitative approaches (Bilsborrow and Okoth Ogondo, 1992; Bingsheng, 1996), and others using quantitative models of micro economics and social psychology.

Rooted in the natural sciences rather than the social, geographers and ecologists have focused on land cover and land use at the macro scale, spatially explicit through remote sensing and GIS, and using macro-properties of social organisation to identify social factors connected to the macro-scale patterns. Due to the poor connections between spatially explicit land use and the social sciences, land-use modellers have a hard time to tap into the rich stock of social science theory and methodology. This is compounded by the ongoing difficulties within the social sciences to interconnect the micro and macro levels of social organisation (Watson, 1978). We will take a closer look at this situation, and discuss possible ways out.

3.1   Micro-level analysis

For social scientists, behaviour is the central topic of study. Social science disciplines and subdisciplines have their preferred levels of analysis and often do not communicate across those levels. For instance, psychologists and sociocultural anthropologists tend to work with individuals and small groups; while sociologists tend to specialise in one level of analysis or another, from individuals to small groups to communities. Farming systems analysis is a form of micro-level land-use research, since it focuses on the single farmer and his/her decisions.

This micro-level focus has a major drawback for the analysis of land-use dynamics. Focusing on one level of analysis, e.g., the individual, is fine, as long as we do not make assumptions or inferences about the other levels of analysis. Unfortunately, assumptions and inferences about other levels are often made, either explicitly or implicitly (Watson, 1978). There are substantive reasons why theories obtained at different levels of analysis often do not match. Human behaviour varies with group size. This is clear from the literature in sociology and social psychology. For example, groups of people will make riskier decisions than individuals; social conformity increases with group size; and inhibitions decrease with group size. In most social studies, we seem to assume that human behaviour is fixed regardless of group size. In this sense, this type of research is reductionist in viewpoint, claiming that individual preferences, decisions, and actions are the fundamental units through which large scale patterns and processes must be explained. By focusing on one level, we restrict our capacity to comprehend: partial analysis may significantly restrict understanding of the interconnections and subtle interrelations among components.

Another characteristic is that social science is generally more concerned with why things happen than where they happen. Relatively few social scientists outside the field of geography value the importance of spatial explicitness, nor do the typical social science data sets contain the geographic co-ordinates that would facilitate linking social science data and remotely sensed or other geographic data (Rindfuss and Stern, 1998). The lack of a spatial perspective in micro-studies ignores the context of the studied behaviour. People live their lives in contexts, and the nature of those contexts structures the way they live (Fotheringham, 2000), while at the same time behaviour influences the spatial configuration of this context. When the individual is the unit of analysis, the individual's household is also a context, as well as the community, the bio-physical environment and the political powers to which the individual might be subjected. Contexts can provide opportunities or present constraints. Hypotheses from theories of context may involve additive effects or interactive effects but in either case, the hypotheses deal with the effects of context on individuals or households (Rindfuss and Stern, 1998). New theories, linking individual behaviour to collective behaviour are being developed to deal with scaling issues in the social sciences and explain emergent (macro) phenomena. Such a meso-level study typically explores how individual people interact to form groups and organise collective action, and how such collective decisions vary with group size, collective social capital, and so on. Game theories are often an important source for explanations at the meso level. Strongly related to land use are well-known studies such as those of Ostrom (1990), that focus on common property management.

At the same time, new research projects attempt to link social science research with geographical data (Geoghegan *et al.*, 1998; Walsh *et al.*, 1999; Walker *et al.*, 2000; Mertens *et al.*, 2000). This type of linkage between socio-economic and geographical data can be a means to provide information on the context that shapes social phenomena.

Only these type of developments can avoid that micro-analyses are carried out in a contextual vacuum and that the analysis destroys the wholeness of the context by limiting the researcher's focus of attention and concept of relevance.

3.2   Macro level analysis

Apart from macroeconomics and some qualitative studies of macrosociology (about the centre-periphery hypothesis, global governance, the 'end of history', etc.), the social sciences are not well developed at the macro level of human organisation. Due to its roots in physical science and system-oriented ecology, however, land-use science has often adopted the macro level approach as its natural style of thinking. Usually, this approach aims to unravel the processes that have caused land-use change, based on statistical analysis of observed patterns of land use, relating these to changes in macro-level variables such as population density, tenure systems, agricultural prices and so on (e.g., de Koning *et al.*, 1998).

If macro level analysis is carried out at the macro scale (which is often, although not necessarily the case), it is able to reveal processes that work on that scale primarily, such as a possible large-scale patterning of intensive-extensive-extractive land use zones around urban centres (e.g., de Groot, 1999). Macro level analysis at the macro scale also appears to be a useful exploratory approach. By working high up in the cone of spatial resolution, macro scale studies may be used as a lens or filter to focus on areas and driving factors that may require more attention, e.g., through identifying the bounds of a complex system and subdividing it into more tractable components.

In macro scale analysis, the measurement of relevant variables is often problematic, in practice. Migration, for instance, may be partly explained by the spatial distribution of economic opportunities. The more valid variable, however, is the perception of these opportunities by potential migrants, mixed with the degree of perceived risk at the potential place of destination which, in its turn, partly depends on the degree to which family or ethnic group members are already settled at that place. Such data are usually outside the reach of macro-scale studies. Note, however, that this problem is not intrinsic. Theoretically, it is perfectly possible to connect macro-scale maps of opportunities with an individual level (micro level) model of migration decision making, in which such factors are incorporated (e.g., de Groot and Kamminga, 1995).

Because macro level approaches most commonly use statistical correlation techniques as their primary tool to quantify the relationships between land-use change and assumed driving factors, they intrinsically suffer from the general weakness of all statistical approaches, namely, the inability to establish causality. The same pattern may be produced by different processes, and a given process can produce different patterns. Sometimes, even the causal direction remains unclear. For instance, does high population density cause low forest cover, or does recently cleared forest cause a high density of people, filling in the empty space?

3.3   Multi scale analysis

Both, the micro level and macro level analysis paradigm have their specific strengths and weaknesses. Both approaches can be greatly improved by including multiple scales. To achieve this we should abandon our natural tendency to associate macro scale studies with macro levels of organisation, and micro scale studies with micro levels of organisation. We are used to associate the concept of ecosystem with something big like a forest or a lake, and we continue to think that macro economics refers to large-scale entities such as nations, only. We can, however, study the ecosystem around a root tip in the soil, just as we can study the macro economics of a village. And reversibly, we can study the individual ('aut-')ecology of the tree that is much larger than its root tip, just as we can study the micro economics of a multinational corporation. This implies that both micro level and macro-level studies can be multi scale.

To start, we may imagine a model structure that is micro level throughout. In other words, it is composed of decision making actors throughout. The model being multi scale, some of these actors will be farmers, others will be national agencies, and others may be global players, but all will be causally interconnected. For their decision making, all actors will look at the variables that concern them at their own level, such as total farm income, soil characteristics, local culture; national food production, national pride, national distribution of population and forest; global relative competitiveness, global biodiversity hotspots, and so on. Many of these variables have a spatial dimension, thus, all actors can be accompanied by the maps that regard them at their level and shape their contexts. In subsection 5.1 we will give an example of a research approach that might form the backbone of this micro level multi scale approach.

However, such an approach remains (multi-)micro, hence causally maybe strong, but never able to explain the emergent system-level (macro) variables, such as prices or forest distribution, that it uses to model the decisions of the actors.

Second, we may imagine a model structure that has its origin in macro level approaches through studying the spatial patterns of land use with the help of GIS and/or remote sensing and relating these patterns to (proximate) variables that represent aggregate processes seen as driving forces (Figure 2.2). Being multi scale, the extent and resolution of analysis are varied from very coarse all the way down to, say, the extreme of one pixel representing approximately one farm. The statistical analysis connecting observed land use to assumed driving factors may then be run at all these scales, from the village (multi farm) scale upwards, each scale connected to its own hypotheses and theories. More importantly, each scale may also be connected to its own micro level auxiliary models, as indicated in the migration example. The basic step is to identify the actors that have the particular map 'in their head' as a co-determinant of their decisions. At national scale, for instance, for the agricultural planning agency the optimum distribution of crop types is important, as is the prospective migrant using an other national-level map, and the logging corporation using yet another. In that way, the 'column' of multi-scale maps becomes covered with actor models like a Christmas tree, each actor model replacing a statistical relationship by a set of essentially verifiable causal assumptions. In subsection 5.2, we will discuss a potential candidate that could be the backbone of such a structure.

Finally, it may be noted that the two approaches micro-level multi-scale analysis and macro level multi scale analysis begin to interweave. A description of this as yet highly imaginary 'multi level land-use change model' is outside the scope of this chapter. It indicates, however, that steps towards a consistent

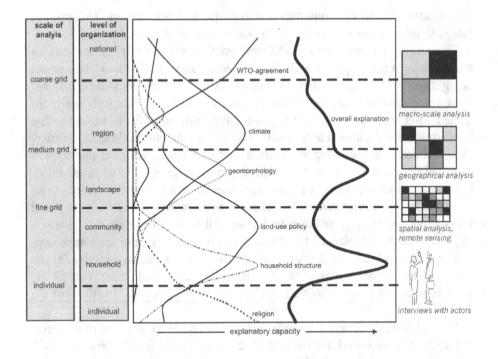

Figure 2.2. Hypothetical magnitude of influence of a number of driving factors on land-use change. For individual driving factors, 'optimal scales' of influence can be distinguished, for the system as a whole there is no, 'optimal scale'.

construction of the two multi scale approaches are steps on a progressive road and that the micro level and macro level approach, originating from different disciplines, may truly complement each other.

4.   Modelling of land-use change

Models of land-use change provide tools for understanding the causes and consequences of rapid land-cover changes. Models are useful for disentangling the complex suite of socio-economic and biophysical factors that influence the rate and spatial pattern of land-use change and in estimating the impacts of changes in land use. Furthermore, models have the capability to make predictions of possible developments of the land-use system in the future. Reviews of deforestation and other land-use change models are provided by Pearce and Brown (1994); Lambin (1997); Kaimowitz and Angelsen (1998) and Bockstael and Irwin (2000). Instead of repeating these reviews, three typical approaches of modelling will be shortly discussed. Because economic models of land-use change make up the largest part of all existing models, a short section will be dedicated to these models. Subsequently, two modelling approaches not based on economic theory, but finding their origin in the micro- and macro-research perspectives mentioned above, are discussed to illustrate the diversity in modelling approaches found.

4.1 Economic models of land-use change

Based on the principles of economic theory, a wide variety of land-use models have been developed, as reviewed by Kaimowitz and Angelsen (1998) and Bockstael and Irwin (2000). A large number of economic land-use change models begin from the viewpoint of individual landowners who make land-use decisions with the objective to maximise expected returns or utility derived from the land, and use economic theory to guide model development, including choice of functional form and explanatory variables. Microeconomic theory is used to provide a rationale for the development of the economic structural equations.

The simplest models are econometric models, explaining land-use change as a function of a number of exogenous factors in one or more regression equations based on transversal or longitudinal data-sets, that may be spatially explicit (e.g., Pfaff, 1999; Jones *et al.*, 1995; Chomitz and Gray 1996). The quantification of this type of models is useful to obtain insights into the influence of different factors on land-use change. The capability of such models to predict future patterns is generally low. Most dynamic models are based on linear and non-linear programming (Bouman *et al.*, 1999; Barbier, 1998). Linear programming models optimise the land-use configuration and management under a number of agro-technical, food security, socio-economic and environmental objectives. Results of a run by a linear programming model are characterised by optimised objective values and the associated optimal set of decision variables. Problems with this type of models include economically non-optimal behaviour and an underestimation of the role of institutions, population, and biophysical factors all of which may play a key role, in reality. In recent efforts to better take into account such factors, constraints have been added to the economic optimisations. Difficulties arise from scaling these models, as they have primarily been designed to work at the micro level. Jansen and Stoorvogel (1998) and Hijmans and van Ittersum (1996) have shown the scale problems that arise when this type of models are used at higher aggregation levels. Potentially, linear programming models could be adapted to take multiple scales into account, essentially in the way sketched in section 3, concerning all micro models. This type of models might lead to good results as long as the assumption that economic processes are the main drivers of land-use change holds.

At the macro level (and usually macro scale), Computable General Equilibrium (CGE) models make certain prices endogenous and thus may be of value in showing how macroeconomic policies and trends affect these prices, several of which may have an effect on land use dynamics. The models themselves are not spatially explicit yet, implying that other models will have to be used in predicting how price changes will affect land use. Due to their high data requirements, CGE models (Fischer and Sun, 2000) might be difficult to make fully operational.

All in all, this overview indicates that economists have been able to make substantial contributions to the understanding of land-use change, in spite of the fact that land-use choices are usually made on a basis that is wider than economics only. Economics will have an important role to play in building the multi-scale edifice of which the principles were designed in the previous section, both from the micro- and macro-level perspective.

4.2   Multi-agent models

Multi-agent models simulate decision making by individual agents involved in land-use change, explicitly addressing interactions among individuals. The explicit attention for interactions between agents makes it possible for this type of models to simulate emergent properties of systems. If the decision rules of the agents are set such that they sufficiently resemble human decision making, they can simulate what we have called the meso level of social organisation in subsection 3.1.

Multi-agent systems are part of distributed artificial intelligence methods. An agent is

> a real or abstract entity that is able to act on itself and on its environment; which can, in a multi-agent universe, communicate with other agents; and whose behaviour is the result of its observations, its knowledge and its interactions with other agents (Sanders *et al.*, 1997).

It can shed light into the degree in which system-level properties simply emerge from local evolutionary forces, and the degree to which those local processes are influenced and shaped by their effect on the persistence and continued functioning of ecosystems or the biosphere (Levin, 1998). Until some years ago mathematical and computational capacity limited the operation of this type of models. Currently, different groups have developed systems to support these kinds of simulations, most often for totally different purposes than land-use change modelling. The High Level Architecture (Lutz, 1997), the Dynamic Interactive Architecture System (DIAS, 1995), and the Multi-modeling Object-Oriented Simulation Environment (Cubert *et al.*, 1997) were developed to support battlefield simulation. The best known system that can be adapted for ecological and land-use simulations is the SWARM environment, developed at the Santa Fe Institute (Hiebler *et al.*, 1994). This type of models will also be suitable for land-use change modelling based on detailed information on socio-economic behaviour under different circumstances. The information about behaviour obtained through extensive field studies of sociologists can be put in context, the relative importance of the different processes influencing land-use change can be tested through sensitivity analysis and a link to higher levels of aggregation can be made: the simulated local-level processes and interactions result in the land-use dynamics at more aggregate levels.

Efforts are currently underway to built operational multi-agent models for land-use change (Vanclay, 1998; Bousquet *et al.*, 1998; Manson, 2000) that adopt this approach. They are, however, pioneering efforts that might guide the direction for this type of modelling.

In multi-agent models, not all agents have to be of the same kind (say, farmers). If, for instance, one group of agents would simulate small farmers and other agents would simulate logging companies, big landowners, a government agency and the World Bank, and if all actors would be interconnected appropriately (e.g., the loggers being attracted to the forest and building a road, the farmers being attracted to the road and migrating, the government subsidising the farmers and not taxing the loggers, and the World Bank threatening to withdraw loans if this is not reversed), a multi-agent model would begin to show the structure of what we will call a multi-scale 'Action-in-Context' model in subsection 5.1.

4.3   Spatially explicit macro-models

While most models focus on 'how' and 'when' land use change occurs, spatially explicit models commonly focus on the 'where' question. So, explicit attention is given to the spatial interconnections between locations, influencing land-use change. Most explicitly this is dealt with in so-called cellular automata models (Balzter *et al.*, 1998; Engelen *et al.*, 1995; Wu, 1998). Cellular automata are a class of mathematical models in which the behaviour of a system is generated by a set of deterministic or probabilistic rules that determine the discrete state of a cell based on the states of neighbouring cells. Despite the simplicity of the transition rules, these models, when simulated over many time periods, often yield complex and highly structured patterns. The transition rules, often based on our understanding of local behaviour, are difficult to quantify and subject to scale-dependencies. New methods are developed to calibrate these transition rules based upon historical data (Candau, 2000). Other spatially explicit simulation models base their transition rules explicitly on an analysis of either historic transitions or spatial variability in present land-use patterns. The LTM model (Brown *et al.*, 2000) uses artificial neural networks to quantify the transition probabilities, whereas the CLUE model (Verburg *et al.*, 1999a), described in more detail below, and GEOMOD2 (Pontius *et al.*, 2001) use multiple regression techniques to quantify the relations between land use and its (proximate) driving factors. These empirical techniques are used to quantify the assumed processes at the level of analysis to avoid upscaling problems when translating local processes to the regional scale.

5.   Examples: two multi-scale research approaches

This section describes two typical approaches, that explicitly address scaling issues through incorporating multiple scales or levels in the analysis.

5.1   Multi-scale micro-level analysis: Action-in-Context

De Groot (1992) has developed a methodological framework for analysis, explanation and solution of environmental problems. Central features of this 'Problem-in-Context' approach are twofold:

(1) the conceptualisation and analysis of environmental problems as empirical/normative discrepancies of facts ('chains of effects') and values ('chains of norms'), that identify the problematic human actions;

(2) the explanation of the environmental problem by setting it in its normative, physical and social context. Solutions are then designed on the basis of options identified in the analysis and explanation.

For the social-scientific explanation of environmental problems, the 'Action-in-Context' (AiC) methodology has been developed as part of the Problem-in-Context framework. Allied, inter alia, to the work of Vayda (1983) and Blaikie (1985), Action-in-Context is a fully micro-level (actor-based) approach that may be put to work also in the explanation of land-use change (de Groot and Kamminga 1995; van den Top, 1998). Action-in-Context starts from the 'primary actions', defined as actions that directly affect the environment, such as farming or logging. It then proceeds by placing these actions in their wider social and physical context, first identifying the 'primary actors' as the social entities that decide on the primary actions, and then connecting these decisions to other (secondary, tertiary, etc.) actors and factors, 'progressively contextualizing', as Vayda (1983) puts it, on a route guided by principles of problem relevance.

In somewhat more detail, the Action-in-Context approach may be described as an interconnection of three structures.

Action, actor, options, motivation Land-use relevant actions (primary or other) are connected to actors, defined as the decision-making social units of these actions, be they individuals or organisations. The decision-making outcome is shaped by two major sets of factors:

(a) the alternative courses of action available to the actor (the actor's options);

(b) the choice-relevant characteristics of these options for the actors, called motivational factors.

For the farmer as an actor, characteristic options consist of crop choices, but may also comprise seasonal or full migration, off-farm work and ('Boserupian') investments in land quality improvements, such as terracing. Motivational factors may be the well-known elements of economic decision-making (cost, benefit, risk, investment level, etc.), but may also comprise factors such as tradition, membership, gender roles, etc. For another land-use relevant actor, such as an agricultural government agency, characteristic options are diffusion of improved varieties, regulation of tenure, subsidies on fertiliser, etc., and characteristic motivational factors are satisfaction of food demands, satisfying the land hunger of poor farmers and big politicians, etc.

Both, land-use options and motivational factors may have a certain overall structure. For the options, these may be called 'strategies', examples of which are area expansion, agricultural transition or (for logging companies), 'grab-it-and-run'. Structures of motivational factors are often called 'actor models'; they may be of the economic kind (i.e., a structure of cost-benefit analysis), but may also follow reasoning closer to, for instance, the ethics of care.

The actors field   Actors are connected to each other. Referred to as 'social networks' or a similar term, social science usually conceptualises connections between them as direct, actor-to-actor, often face-to-face, linkages. In that conception, a farmer is primarily connected to other farmers (neighbours, family, friends, etc.). Such social networks are of obvious value in the study of culture, village politics and many more issues, but at the same time, it is questionable whether such linkages really explain land use. As an example, we may imagine that the village consists of migrants who recently settled along a logging road. Then, the land use of all the farmers is explained through a very different actor than the farmers themselves, namely, the logging company. By cutting all the big trees, the logging company created the option, and by building the road, the company created the connection to the market, hence the favourable farm-gate prices. In other words, for each farmer, the actor of influence, that is to say the actor that explains their land use, is the logging company, irrespective whether the company is inside or outside of their social network.

Hence, causal actor linkages are of a different kind than social networks. Causal social linkages are linkages of power, and having power over other people is not primarily related to frequent contact. Primarily, power is having influence on other people's choices, which implies having influence on other people's options and/or motivational factors. Some examples from the deforestation field are: government agencies opening up new lands (= creating options for the poor), government agencies issuing subsidies for forest ranching (= influencing motivational factors of the rich), landlords prohibiting tenants

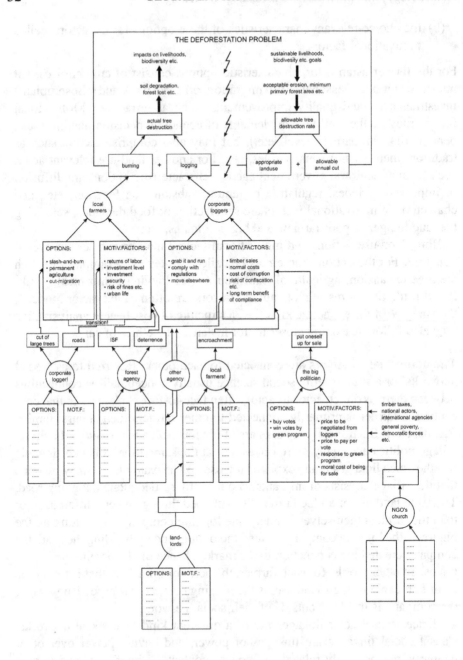

Figure 2.3. An example of an actors field of deforestation in the Sierra Madre region, Philippines. After de Groot and Kamminga (1995).

to plant trees (= closing off an option), consumers becoming reluctant to buy tropical timber (= influencing motivational factors of logging corporations), and so on. The social-scientific explanation of what actors are doing should follow these 'power lines' between actors. On the progressively contextualizing route, the researcher first identifies the actors physically interacting with the environment (the 'primary actors'), then their options and motivational factors, then the 'secondary' actors influencing these options and motivational factors, and so on. In the Action-in-Context framework, actors field is the term given to the structure found in that way. Figure 2.3 shows an example of an actors field of deforestation in the Philippines (de Groot and Kamminga, 1995). The different actors identified within the actors field often operate at different scales: e.g., primary actors that physically interact with the environment cause land-use change at the spatial scale of the farm and surrounding area. At the same time, higher-level organisations, such as governments and institutions such as the World Bank that act as secondary or tertiary actors, operate at spatial scales up to the national and global extent. Note that through constructing roads, loggers are secondary actors to the farmers and through encroachment on concessions, farmers are secondary actors to the loggers. The link between the loggers and the big politician summarises the corruption market; driven by the need for cash (to buy votes), the politician sells the powers of his public mandate for private profit.

The deeper structure model An actors field model is typically construed by hopping over, as it were, to a next actor category as soon as options and motivational factors of one actor category have been identified. For each actor category separately, however, its options and motivational factors may also be seen as causally connected to underlying properties of the structure and culture in which the actor is embedded. As an example from the field of land use, the option to build a terrace is a real option for a farmer only if he knows how to build it or, in case he does not, if he has the social, financial or other capital to tap the knowledge of the local expert. Even if the knowledge is provided, he still needs capital for the investment, either by doing it himself and thus foregoing current income, or by paying others. Thus, knowledge and capital are two factors determining the actual options for a farmer.

This full picture is given in Figure 2.4. At the top, one finds the rectangle of action/actor/options/motivations that is also the repeated core element of Figure 2.3. Underlying the options are the two factors discussed already, called here 'potential options' and 'autonomy'. Potential options are defined as anything the actor could do if he were infinitely autonomous ('rich'). Autonomy is defined as all resources the actor has access to (financial, social, etc. capital), but restricted by effective regulations, taboos and other prohibitions. Underlying the motivational factors are two factors given in Figure 2.3, one called 'objectified motivations' and the other called 'interpretations'. Objectified motivations are defined as all decision-relevant characteristics of options that are easily quantifiable, such as economic costs and benefits, hours of transport time, caloric value of food, and so forth. Interpretations then, are all 'multipliers' attached to the objective factors by the actor, to form motivational factors as perceived and valued by the actor. One causal level below these, the four factors are seen to be determined by 'micro-structure', which is roughly the meso level of social organisation, and the macro level, defined as all structures outside the influence of the actor himself (e.g., markets or political structure).

Figure 2.4 summarises common sense logic in a way that improves well-known decision models of social psychology. It also interconnects contributions from several other social science disciplines. The 'potential options' element is the natural focal area of agronomy and other technical disciplines; roughly, agricultural extension brings new potential options to farmers. The 'autonomy' element comprises the economic attention to endowments, but also the sociological interest in (private) social and cultural capital and empowerment. The 'objectified motivations' element is the area where microeconomists find their home, the 'interpretations/frames/world views' element covers many of the contributions from anthropology, psychology and farming styles research, 'micro-structure' is a focal point of institutional economists, studies on (collective) social capital, geography, political arenas, etc., and 'macro-structure' is the meeting point of macroeconomics and sociology.

The environment is part of the micro- and macro- structure. 'micro- environment' represents the small-scale patterns of soils, etc., that determine options, costs and benefits at the next-higher causal level, and 'macro-environment' represents the large-scale systems that influence land-use decisions either directly (such as climate) or indirectly, through the micro-structure. Many elements in macro-structure are non-spatial scalars, such as most prices emerging from markets. Many others are, however, spatially defined, such as the environmental factors, but also socio-economic factors, such as the spatial differentiation of expected income and risk connected to migration. Essentially, what should be mapped in a model are all the elements that are mentally mapped by the actor as the basis for his decision. This holds for all types of actors. Thus, maps in the farmer's model will be different (and often smaller-scale) than maps of actors such as a land bank or an environmental agency.

Applying and computerising Action-in-Context   For directly policy-relevant applications, a strength of AiC is that the actor categories identified in the actors field coincide with potential target groups of policies, while the deeper structure model identifies options for policy content. This direct connection between empirical analysis and policy design follows from AiC's causal focus. For such directly applied research, or as qualitative studies, it is not necessary

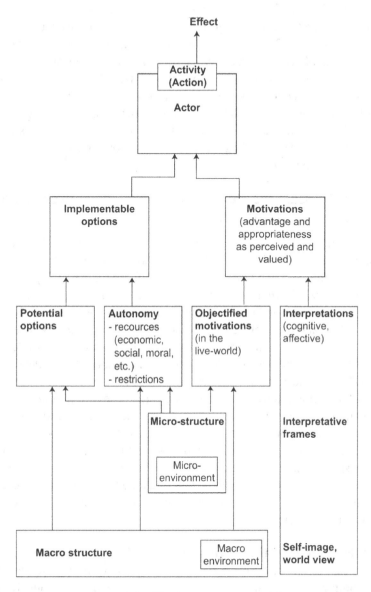

Figure 2.4. The deeper structure model.

to formalise AiC into a quantified and fully interconnected model. A natural course of action, for instance, is to first identify the actors field and then focus on a deeper structure analysis for the most relevant actors only.

Due to its relatively rigid, simple and logically coherent principles, Action-in-Context may also be formalised and computerised as a single, quantifiable model, using, for instance, one of the environments created for multi-agent modelling, as discussed in the preceding section. Linkages between actors should then not run from the actions of one actor to options and/or motivations of another (as in Figure 2.3), but to the level of 'potential options/autonomy/...' and to the level of 'micro-structure/macro-structure' of the deeper structure model (Figure 2.4). A decision by a government agency to open up a new area of forest to logging concessions would then, for instance, change the map of potentially available forest (a part of 'micro-structure') of the logging company actors. In that way, the actors field and the deeper structure models become united in a single whole, including the multi-scale maps connected to the various categories of actors in the actors field.

5.2   Multi-scale macro-level analysis: the conversion of land use and its effects

The Conversion of Land Use and its Effects or CLUE methodology (Veldkamp and Fresco 1996; Verburg *et al.*, 1999a) is based on theories on the functioning of the land-use system, derived from landscape ecology (Holling, 1992; Levin, 1992; Turner and Gardner 1992). To a certain extent, natural ecosystems correspond in structure, function and change to land-use systems (Conway, 1987) and the social systems underlying changes in land use. Social systems and agro-ecosystems are, just like natural ecosystems, complex adaptive systems that can be described by theories and methodologies developed in ecology (Holling and Sanderson, 1996; Adger, 1999; Levin *et al.*, 1998).

The CLUE methodology consists of two parts. The first part aims at establishing relations between land use and its driving factors, explicitly taking scale dependencies into account. The second part aims at dynamic modelling. The multi-scale analysis of the driving factors of land-use change is based on the analysis of spatial patterns of actual land use. Except for areas with minimal human influence, these patterns are the result of a long history of land-use change and contain, therefore, valuable information about the relations between land use and its driving factors.

Multi-scale analysis of the driving factors of land-use change   Because it is assumed that the relations between land use and driving factors are extremely complex due to scale dependencies, interconnections and feedback mechanisms, no attempts are made to unravel the individual processes. Instead, empirical relations between land use and its supposed determining factors are used to explain the observed pattern of land use, e.g., through regression analysis. Another characteristic of the approach is that no a priori levels of analysis, e.g., landscape or regional level, are superimposed. Instead, the analysis is repeated at a selection of artificial resolutions, imposed by gridded data structure.

This method was used to study the (proximate) driving factors that determine the spatial distribution of land use in a number of case studies. The case study in China is used here as illustration (Verburg and Chen, 2000). These driving factors were studied by systematically varying the spatial extent and resolution of analysis (Figure 2.5). The extent is varied between the country as a whole, and a subdivision in eight regions. The resolution of analysis was increased in six steps from grid-cells of 32×32 km^2 ($\sim$1000 km^2; n=9204) to grid-cells of 192×192 km^2 ($\sim$36800 km^2; n=258). For these different scales of observation, the land-use distribution (represented by the relative area covered by the individual land-use types in the grid cells) was related to a large set of potential driving factors by correlation and regression analysis. As an example of the results (reported by Verburg and Chen (2000)), the five explanatory factors having the highest correlation with the distribution of cultivated land are presented in Table 2.1. The analysis demonstrated that for the country as a whole the pattern of cultivated land corresponds best to the agricultural population distribution. When the extent of analysis is reduced, by repeating the statistical analysis for separate regions within China, agricultural population is not always the most important explanatory variable.

The effect of resolution was studied in a similar way. Figure 2.6 displays the correlation coefficients between a selection of variables and the cultivated land area at the six aggregation levels. For all variables the correlation coefficient increases with grain size. This increase in correlation coefficient may be the result of area aggregation, which reduces variability. However, the correlation coefficients were not inflated consistently with successive aggregation and for the different variables.

Whereas the correlation coefficients of maximum temperature and mean elevation remain approximately constant over the range of grain sizes, the coefficients for soil suitability and urban population density strongly increase with grain size. The change in correlation structure is related to the spatial variability of the variable and the distance over which the parameter affects land use. The strong increase in correlation between urban population and cultivated land can be explained by the influence of a city on land use in the surrounding area. With the increase in grain size, an increasing part of the cultivated land around a city falls within the same grid-cell as the urban population, yielding higher correlation.

Also for other countries for which a similar analysis was made, a considerable influence of spatial scale (resolution and extent) on the identified driving factors was found (Veldkamp and Fresco, 1997; de Koning *et al.*, 1998; Verburg and Chen, 2000; Kok and Veldkamp, 2001). The results for the different case studies show that the relations between land use and its predictors are indeed influenced by the grain and extent of analysis. The methodology applied enables

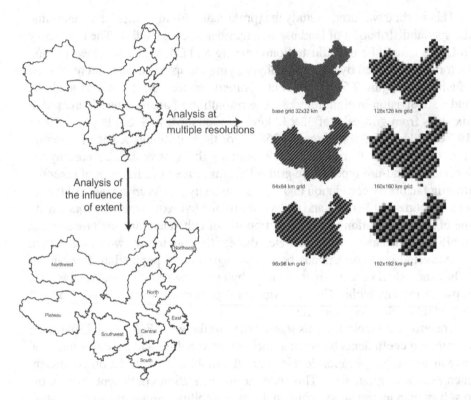

Figure 2.5. Subdivision of China into different units of extent and resolution to allow analysis at multiple scales.

identification of these scale dependencies. However, it is not possible to exactly unravel the processes explaining these scaling properties. Possible explanations for the influence of the spatial scale of analysis are:

- The reduction in spatial variability: coarse grain sizes obscure variability, while fine grain sizes obscure general trends. Shifts in grain size may produce more than averages or constants: they may transform homogeneity into heterogeneity and vice versa (Kolasa and Rollo, 1991).

- Emergent properties: changes in grain size are frequently associated with new or emergent properties. In complex, constitutive hierarchies, characteristics of larger units are not simple combinations of attributes of smaller units.

- The influence some factors can have over a considerable distance. At coarse grains, these factors fall within the same unit of analysis and cause therefore a change in correlation structure.

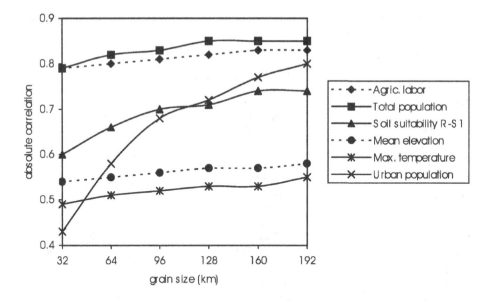

Figure 2.6. Absolute correlation between the relative cover with cultivated land and a number of explanatory factors at different resolutions for China; all correlations significant at P<0.0001; source: Verburg and Chen, 2000.

- Stronger overlap among variables. Aggregation reduces intra-class variance and the size of the sample population, smoothing the distribution and reducing the number of outlier values identified within each class. This can create strong overlap among variables, greatly reducing the potential value of such variables for distinguishing classes.

- The influence of the extent of analysis can be explained by the decreasing importance of local situations with an increasing extent of analysis. Our analysis for the different regions has shown that relations found at the national extent are not always valid at the regional extent and vice versa. A smaller extent also allows introduction of specific variables that are important for the area under analysis. So, a smaller extent offers better insight into the specific situation of the region, while a larger extent allows identification of the general patterns. At the same time, it often holds that the larger the region under study, the larger the proportion of the interactions that are internalised, other things being equal (Wilbanks and Kates, 1999).

Multi-scale simulations The empirical relations at multiple scales derived from the scale sensitive analysis are directly used in a dynamic model. The dynamic model has the following characteristics:

Table 2.1. Most important explanatory factors for the distribution of cultivated land based on Pearson correlation coefficients*.

Whole country (n=9204)	Northeast (n=764)	North (n=674)
Agric. population	Max. temperature	Mountains
Agric. labour	Mountains	Agric. population
Total population	Deep soils	Slope
Rural labour	Level land	Agric. labour
Soil suitability R-S1	Agric. population	Plain land
Northwest (n=3361)	East (n=336)	Central (n=554)
Agric. population	Total Precipitation	Mountains
Agric. Labour	Mountains	Agric. population
Rural labour	Elevation range	Total population
Total population	Well drained soils	Plain land
Loess	Bad drained soils	Well drained soils
South (n=547)	Southwest (n =1098)	Plateau (n=1870)
Mountains	Agric. labour	Rural labour
Slope	Agric. population	Agric. labour
Max. temperature	Rural labour	Agric. population
Avg. temperature	Total population	Total population
Mean elevation	Mean Elevation	Urban population

* all significant at $P < 0.0001$

- All simulations are executed in a spatially explicit way, so that the geographical pattern of land use change results.

- Allocation of land use changes is based on the dynamic simulation of competition among different land use types. Competitive advantage is based on the 'local' and 'regional' suitability of the location and the national level demand for land use type-related products (e.g., food demand or demand for residential area).

- The 'local' and 'regional' suitability for the different land use types is determined by the quantified relations between land-use and a large number of explanatory factors as presented above.

- Different scenarios of developments in land use can be simulated. At the national level, scenarios include different developments in agricultural demands that can be determined on the basis of developments in consumption patterns, demographic characteristics, land use policies and export volumes. At the sub-national level, different restrictions towards the allocation of land use change can be implemented, e.g., the protection of nature reserves or land allocation restrictions in areas susceptible to land degradation.

In China, land use change has recently been dominated by losses of agricultural land due to urbanisation and desertification. A scenario was defined at the national level, taking into account the major issues affecting Chinese land use: increasing population size and structure, growing food demand, increase in urban built-up areas, reforestation and desertification (Verburg *et al.*, 1999b). Spatially explicit land-use changes in this scenario were simulated for the period 1991-2010. The model results, expressed as changes over the simulated period for two land use types: cultivated area (mainly arable crops) and unused area (mainly desert and wasteland) are shown in Figure 2.7. Changes are biggest in the western part of the agricultural area, located in the centre of the country. For these areas, the model results indicate that, while substantial parts of formerly cultivated land become unusable because of land degradation, other parts are being reforested. Cultivated land is also lost to increasing urban areas in Eastern and Southern China. This land has a higher productivity than average, so these area losses have relatively stronger consequences for total crop production volumes.

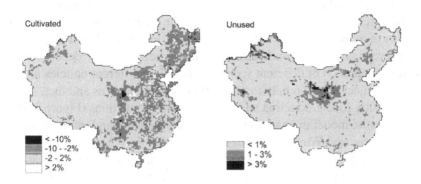

Figure 2.7. Land use change in China for cultivated land and unused land (mainly desert) during 1991-2010 simulated by the CLUE model for a baseline scenario.

6.   Discussion: applications of land-use change knowledge and challenges for future research

Apart from the intellectual challenge to understand the land use system and its complexities, there is a great need for spatially explicit assessments of the dynamics of the land-use system. Land-use system information is essential to assess the effects of human influence on natural resources. Estimates of carbon emissions are subject to high degrees of uncertainty, due in large part to difficulties in assessments of land conversion across local to regional scales (Dietz and Rosa, 1997; Moran, 2000). It is also a consequence of the fact that terrestrial vegetation and soils are extremely heterogeneous over the land surface, and estimates of the magnitude of carbon emissions during clearing and of carbon sequestration during regrowth of vegetation vary substantially (Fearnside and Embrozia, 1998; Hughes *et al.*, 2000). Hence, for improved assessments there is a need for spatially explicit assessments of land-use change. Similar considerations hold for assessments of changes in nutrient balances, important for sustainability in agricultural production and environmental quality (Smaling and Fresco, 1993; de Koning *et al.*, 1999). Other applications that directly use spatially explicit information on land-use change are erosion/sedimentation models (Schoorl and Veldkamp, 2001), water resource assessments (Wear *et al.*, 1998) and biodiversity assessments (Lebel and Murdiyarso, 1998; Chapin *et al.*, 2000; van der Meer *et al.*, 1998). Maintaining biological diversity depends on the spatial arrangement of land uses in the landscape, because of the effects of fragmentation and habitat size on the risks of extinction. The spatial arrangement of land uses also has implications for ecosystem function, and hence its goods and services. For example, the way in which residual forest is distributed, relative to the stream margin can have an impact on soil erosion and hydrology of a catchment. Finally, how natural disturbances like fires, pests and diseases propagate through a heterogeneous landscape clearly depends on its structure. Evaluation of different scenarios, based on different policies and development trajectories can help to assess these spatial patterns and their effects on biodiversity, food production and other ecosystem functions (Figure 2.8). In this way, spatial modelling can improve land-use planning and informed policy making (Farrow and Winograd, 2001).

Apart from these biophysical assessments, land-use change studies can also help to identify vulnerable people and places in the face of global change. Differential impacts and abilities to respond create winners and losers in this change. Whether involving land fragmentation, degradation of agricultural productivity, declines in economic well-being, or involuntary human migration, land-use/land-cover change plays an important role. An example of such an effect of land-use change on social behaviour is found in research by Stephen Walsh and colleagues in Northeast Thailand (Walsh *et al.*, 1999). Here it is suggested that young adults decide to out-migrate when the land is highly fragmented, suggesting that limited availability of land and other resources act as a feedback on human behaviour. The results of spatially explicit land-use change models allow straightforward identification of areas that are likely to face high rates of land-use change in the near future, so-called 'hot-spots' of land-use change. Based upon results of the CLUE modelling framework, Verburg *et al.*, (2000) identified regions in China that are lagging behind in economic development while, at the same time, these regions are faced with deteriorating land resources. Identification and understanding of the dynamics underlying these regional differences will help to focus in-depth research and policy intervention at the most threatened regions.

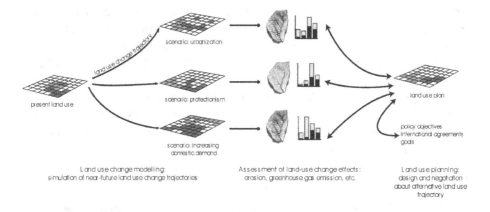

Figure 2.8. The role of land-use change modelling within studies aiming at improved land use planning.

After all intellectual efforts of properly describing land-use dynamics and their effects, the main task scientists are faced with is communication of their results to the stakeholders. Ideally, stakeholders have been involved during the research itself in 'social learning' processes (Röling and Maarleveld, 1999) which ensures that the relevant questions are answered. Appropriate presentation of the results is extremely important. Most stakeholders are not in a position to read scientific papers or bulk reports. For the presentation of spatially explicit assessments of land-use change, the researcher can make use of the visual capabilities of geographical information systems. The presentation of maps to decision-makers is appropriate to communicate results and provoke discussions between policy makers and scientist on the probability of the foreseen changes (Goodchild, 2000). Spatially explicit representation of land-use changes has also proven to be an appropriate means to discuss resource bases, spatial interconnectivities between areas and the consequences of local actions with farmers (Gonzalez, 2000).

The different approaches for land-use change research described in this chapter illustrate differences in world view that underlie how people explain the functioning of social systems (Zimmerer, 1991). Our review of approaches and their application has shown that no single research appraoch is able to answer all LUCC questions. Blending processes and structures at several scales and including their interactions should become the norm in land-use change models, because of the recognition that land-use dynamics originate from the interactions of processes and structures at scales ranging from the individual tree to the patch, region, and even globe. A pluralism of emphases, from individual-based to regional/global models, will continue to be useful for addressing problems at multiple scales, with meta-modelling used when linkage is needed (Baker and Mladenoff, 1999). In doing this, we no longer perceive scale dependencies as a nuisance complicating our research, but rather as a source of information to better understand the functioning of the complex systems under analysis. To achieve this is a true challenge and requires researchers to step beyond their disciplinary traditions (Wilbanks and Kates, 1999). A number of approaches, of which a few are described in this paper, are already available that, from their own discipline try to move beyond the disciplinary boundaries and traditional rigid levels of analysis. These developments will not only benefit our understanding of the land-use system itself, but also add to the study of other complex interdisciplinary systems.

References

Adger N.W. (1999) Evolution of economy and environment: an application to land use in lowland Vietnam. Ecological Economics 31, 365–379.

Allen T.F.H. and Starr T.B. (1982) Hierarchy: Perspectives for Ecological Complexity. University of Chicago Press, Chicago.

Baker W.L. and Mladenoff D.J. (1999) Progress and future directions in spatial modeling of forest landscapes. In: Mladenoff D.J., Baker W.L. (eds). Spatial modeling of forest landscape change: approaches and applications. Cambridge University Press, Cambridge, 333–349.

Balzter H., Braun P.W. and Köhler W. (1998) Cellular automata models for vegetation dynamics. Ecological Modelling 107, 113–125.

Barbier B. (1998) Induced innovation and land degradation: Results from a bioeconomic model of a village in West Africa. Agricultural Economics 19, 15–25.

Bilsborrow R.E. and Okoth Ogondo H.W.O. (1992) Population-driven changes in land use in developing countries. Ambio 21, 37–45.

Bingsheng K. (1996) Regional inequality in rural development. In: Garnaut R., Shutian G., Guonan M. (eds). The third revolution in the Chinese countryside. Cambridge University Press, Cambridge, 245–258.

Blaikie P.M. (1985) The political economy of soil erosion in developing countries. Longman, Harlow (UK).

Bockstael N.E. and Irwin E.G. (2000) Economics and the Land Use-Environment Link. In: Folmer H., Tietenberg T. (eds). The International Yearbook of Environmental and Resource Economics 1999/2000. Edward Elgar Publishing.

Bouman B.A.M. Jansen H., Schipper R.A., Nieuwenhuyse H. and Bouma J. (1999) A framework for integrated biophysical and economic land use analysis at different scales. Agriculture, Ecosystems and Environment 75, 55–73.

Bousquet F., Bakam I., Proton H. and Le Page C. (1998) Cormas: common-pool resources and multiagent systems. Lecture Notes in Artificial Intelligence 1416, 826–837.

Brown D.G., Pijanowski B.C. and Duh J.D. (2000) Modeling the relationships between land use and land cover on private lands in the Upper Midwest, USA. Journal of Environmental Management 59, 247–263.

Candau J. (2000) Calibrating a cellular automaton model of urban growth in a timely manner. In: Parks B.O., Clarke K.M., Crane M.P. (eds). Proceedings of the 4th international conference on integrating geographic information systems and environmental modeling: problems, prospects, and needs for research; 2000, Sep 2-8. University of Colorado, Boulder.

Chapin F.S., Zavaleta E.S., Eviner V.T., Naylor R.L., Vitousek P.M., Reynolds H.L., Hooper D.U., Lavorel S., Sala O.E., Hobbie S.E., Mack M.C. and Diaz S. (2000) Consequences of changing biodiversity. Nature 405, 234–242.

Chomitz K.M. and Gray D.A. (1996) Roads, Land Use, and Deforestation: A spatial model applied to Belize. The World Bank Economic Review 10, 487–512.

Conway G.R. (1987) The properties of agro-ecosystems. Agricultural Systems 24, 95–117.

Cubert R.M., Goktekin T., Fishwick P.A. (1997) MOOSE: architecture of an object-oriented multimodeling simulation system.Proceedings of Enabling Technology for Simulation Science. Orlando Society of Photo-optical Instrumentation Engineers, Orlando, Florida.

de Groot W.T. (1992) Environmental science theory; Concepts and methods in a one-world problem oriented paradigm. Elsevier Science Publishers, Amsterdam and New York.

de Groot W.T. (1999) Van Vriend naar Vijand naar Verslagene en Verder, Een evolutionair perspectief op de verhouding tussen mens en natuur. Nijmegen University Press, Nijmegen.

de Groot, W. T. and Kamminga, E. M. (1995) Forest, People, Government: A policy-oriented analysis of the social dynamics of tropical deforestation. 410100117. Bilthoven, Dutch National Programme on Global Air Pollution and Climate Change.

de Koning G.H.J., Veldkamp A. and Fresco L.O. (1998) Land use in Ecuador: a statistical analysis at different aggregation levels. Agriculture, Ecosystems and Environment 70, 321–247.

de Koning G.H.J., Veldkamp A. and Fresco L.O. (1999) Exploring changes in Ecuadorian land use for food production and their effects on natural resources. Journal of Environmental Management 57, 221–237.

DIAS. (1995) The Dynamic Information Architecture System: a High Level Architecture for Modeling and Simulation. Advanced Computer Applications Center, Argonne National Laboratory http://www.dis.anl.gov/DEEM/DIAS/diaswp.html.

Dietz T., and Rosa E.A. (1997) Effects of Population and Affluence on Carbon Dioxide Emissions. Proceedings of the National Academy of Sciences 94, 175–179.

Dovers S.R. (1995) A framework for scaling and framing policy problems in sustainability. Ecological Economics 12, 93–106.

Engelen G., White R., Uljee I. and Drazan P. (1995) Using Cellular Automata for Integrated Modelling of Socio-Environmental Systems. Environmental Monitoring and Assessment 34, 203–214.

Farrow A. and Winograd M. (2001) Land Use Modelling at the Regional Scale: an input to Rural Sustainability Indicators for Central America. Agriculture, Ecosystems and Environment 85, 249–268.

Fearnside P.M. and Imbrozia R. (1998) Soil carbon changes from conversion of forest to pasture in Brazilian Amazonia. Forest Ecology and Management 108, 147–166.

Fischer G. and Sun L.X. (2001) Model based analysis of future land-use development in China. Agriculture, Ecosystems and Environment 85, 163–176.

Fotheringham A.S. (2000) Context-dependent spatial analysis: A role for GIS? Journal of Geographical Systems 2, 71–76.

Fresco L.O. (1994) Imaginable futures, a contribution to thinking about land use planning. In: Fresco L.O., Stroosnijder L., Bouma J., van Keulen H. (eds). The future of the land: Mobilising and Integrating Knowledge for Land Use Options. John Wiley and Sons, Chichester, 1–8.

Gardner R.H. (1998) Pattern, process, and the analysis of spatial scales. In: Peterson D.L., Parker V.T. (eds). Ecological Scale: Theory and Applications. Columbia University Press, New York, 17–34.

Geoghegan J., Pritchard J.R.L., Ogneva-Himmelberger Y., Chowdhury R.R., Sanderson S. and Turner II B.L. (1998) 'Socializing the Pixel' and 'Pixelizing the Social' in Land-Use and Land-Cover Change. In: Liverman D., Moran E.F., Rindfuss R.R., Stern P.C. (eds). People and Pixels: Linking Remote Sensing and Social Science. National Academy Press, Washington.

Gibson C.C., Ostrom E. and Anh T.K. (2000) The concept of scale and the human dimensions of global change: a survey. Ecological Economics 32, 217–239.

Gonzalez, R. (2000) Platforms and Terraces. GIS as a Tool for Interactive Land Use Negotiation. Wageningen and Enschede, Wageningen University and ITC. PhD thesis.

Goodchild M.F. (2000) The current status of GIS and spatial analysis. Journal of Geographical Systems 2, 5–10.

Goodwin B.J., Fahrig L. (1998) Spatial Scaling and Animal Population Dynamics. In: Peterson D.L., Parker V.T. (eds). Ecological Scale: Theory and Applications. Columbia University Press, New York, 193–206.

Hiebler D., Strom M., Daniel T.C. (1994) The SWARM Simulation System and Individual-based Modeling. Decision Support 2001: 17th Annual Geographic Information Seminar and Resource Technology '94 Symposium, Toronto, 474–494.

Hijmans R.J. and Van Ittersum M.K. (1996) Aggregation of spatial units in linear programming models to explore land use options. Netherlands Journal of Agricultural Science 44, 145–162.

Holling C.S. (1992) Cross-scale morphology, geometry, and dynamics of ecosystems. Ecological Monographs 62, 447–502.

Holling C.S. and Sanderson S. (1996) Dynamics of (Dis)harmony in Ecological and Social Systems. In: Hanna S., Folke C., Mäler K.G., Jansson A. (eds). Rights to nature: Ecological, economic, cultural, and political principles of institutions for the environment. Island Press, Washington, 57–85.

Hughes R.F., Kauffman J.B. and Cummings D.L. (2000) Fire in the Brazilian Amazon. 3. Dynamics of Biomass, C, and Nutrient Pools in Regenerating Forests. Oecologia 124, 574–588.

Intergovernmental Panel on Climate Change, IPCC. (1997) Revised 1996 IPCC Guidelines for National Greenhouse Gas Inventories. Workbook. Paris, OECD.

Jansen H. and Stoorvogel J.J. (1998) Quantification of Aggregation Bias in Regional Agricultural Land Use Models: Application to Guacimo County, Costa Rica. Agricultural Systems 58, 417–439.

Jones D.W., Dale V.H., Beuchamp J.J., Pedlowske M.A. and O'Neill R.V. (1995) Farming in Rondonia. Resource and Energy Economics 17, 155–188.

Kaimowitz, D. and Angelsen, A. (1998) Economic Models of Tropical Deforestation - A Review. Bogor, Center for International Forestry Research.

Kok K. and Veldkamp A. (2001) Evaluating impact of spatial scales on land use pattern analysis in Central America. Agriculture, Ecosystems and Environment 85, 205–221.

Kolasa J. and Rollo C.D. (1991) Introduction: The Heterogeneity of Heterogeneity - a Glossary. In: Kolasa J., Pickett S.T.A. (eds). Ecological Heterogeneity. Springer Verlag, New York.

Lambin E.F. (1997) Modelling and monitoring land-cover change processes in tropical regions. Progress in Physical Geography 21, 375–393.

Lambin, E. F., Baulies, X., Bockstael, N. E., Fischer, G., Krug, T., Leemans, R., Moran, E. F., Rindfuss, R. R., Sato, Y., Skole, D., Turner II, B. L, and Vogel, C. (2000a) Land-Use and Land-Cover Change (LUCC), Implementation Strategy. IGBP Report 48, IHDP Report 10. Stockholm, Bonn, IGBP, IHDP.

Lambin E.F., Turner II B.L., Geist H.J., Agbola S.B., Angelsen A., Bruce J.W., Coomes O., Dirzo R., Fischer G., Folke C., George P.S., Homewood K., Imbernon J., Leemans R., Li X.B., Moran E.F., Mortimore M., Ramakrishnan P.S., Richards J.F., Skanes H., Stone G.D., Svedin U., Veldkamp A., Vogel C. and Xu J.C. (2000b) The Causes of Land-Use and Land-Cover Change: Moving Beyond the Myths. LUCC discussion paper.

Lebel, L. and Murdiyarso, D. (1998) Modelling Global Change Impacts on Tropical Landscapes and Biodiversity. 5. Bogor, BIOTROP-GCTE/ Impacts Centre for Southeast Asia (IC-SEA). IC-SEA Report.

Levin S.A. (1992) The problem of pattern and scale in ecology. Ecology 73, 1943–1967.

Levin S.A. (1998) Ecosystems and the Biosphere as Complex Adaptive Systems. Ecosystems 1, 431–436.

Levin S.A., Barrett S., Aniyar S., Baumol W., Bliss C., Bolin B., Dasgupta P., Ehrlich P., Folke C., Gren I.M., Holling C.S., Jansson A.M., Jansson B.O., Mäler K.G., Martin D., Perrings C. and Sheshinski E. (1998) Resilience in natural and socio-economic systems. Environment and Development Economics 3, 222–235.

Lutz R. (1997) HLA Object Model Development: A Process View, Simulation Interoperability Workshop. Simulation Interoperability Standards Organization, Orlando http://siso.sc.ist.ucf.edu/siw/97spring/ papers/010.pdf.

Manson S.M. (2000) Agent-based dynamic spatial simulation of land-use/cover change in the Yucatán peninsula, Mexico. In: Parks B.O., Clarke K.M., Crane M.P. (eds). Proceedings of the 4th international conference on integrating geographic information systems and environmental modeling: problems, prospects, and needs for research; 2000 Sep 2-8; Boulder. University of Colorado, Boulder.

Meentemeyer V. (1989) Geographical perspectives of space, time, and scale. Landscape Ecology 3, 163–173.

Mertens B., Sunderlin W.D., Ndoye O. and Lambin E.F. (2000) Impact of Macroeconomic Change on Deforestation in South Cameroon: Integration of Household Survey and Remotely-Sensed Data. World Development 28, 983–999.

Moran E.F. (2000) Progress in the last ten years in the study of land use/cover change and the outlook for the next decade. In: Diekmann A., Dietz T.,

Jaeger C.C., Rosa E.A. (eds). Studying the Human Dimensions of Global Environmental Change. MIT Press, Cambridge, Massachetts.

O'Neill R.V., Johnson A.R. and King A.W. (1989) A hierarchical framework for the analysis of scale. Landscape Ecology 3, 193–205.

O'Neill R.V. and King A.W. (1998) Homage to St. Michael; or, Why Are There So Many Books on Scale? In: Peterson D.L., Parker V.T. (eds). Ecological Scale: Theory and Applications. Columbia University Press, New York, 3–16.

Ojima D.S., Galvin K.A. and Turner II B.L. (1994) The global impact of land-use change. BioScience 44, 300–304.

Ostrom E. (1990) Governing the Commons: The Evolution of Institutions for Collective Action. Cambridge University Press, New York.

Pearce D. and Brown K. (1994) Saving the world's tropical forests. In: Brown K., Pearce D. (eds). The causes of tropical deforestation, the economic and statistical analysis of factors giving rise to the loss of tropical forests. University College London Press, London.

Pfaff A.S.P. (1999) What Drives Deforestation in the Brazilian Amazon? Evidence from Satellite and Socioeconomic Data. Journal of Environmental Economics and Management 37, 25–43.

Pontius R.G., Cornell J.D. and Hall C.A.S. (2001) Modeling the spatial pattern of land-use change with GEOMOD2: Application and validation. Agriculture, Ecosystems and Environment 85, 191–203.

Rindfuss R.R. and Stern P.C. (1998) Linking Remote Sensing and Social Science: The Need and the Challenges. In: Liverman D., Moran E.F., Rindfuss R.R., Stern P.C. (eds). People and Pixels: Linking Remote Sensing and Social Science. National Academy Press, Washington.

Röling N. and Maarleveld N. (1999) Facing strategic narratives: An argument for interactive effectiveness. Agriculture and Human Values 16, 295–308.

Sanders L., Pumain D., Mathian H., Guerin-Pace F. and Bura S. (1997) SIMPOP: a multi-agent system for the study of urbanism. Environment and Planning B 24, 287–305.

Schoorl J.M. and Veldkamp A. (2001) Linking land use and landscape process modelling: a case study for the Alora region (South Spain). Agriculture, Ecosystems and Environment 85, 281–292.

Smaling E.M.A. and Fresco L.O. (1993) A decision-support model for monitoring nutrient balances under agricultural land use (NUTMON). Geoderma 60, 235–256.

Turner II B.L. (1994) Local faces, global flows: the role of land use and land cover in global environmental change. Land degradation and rehabilitation 5, 71–78.

Turner II B.L. (1997) Spirals, Bridges and Tunnels: Engaging Human– Environment Perspectives in Geography. Ecumene 4, 196–217.

Turner II B.L. and Brush S.B. (1987) Comparative farming systems. Guilford Press, New York.

Turner II, B.L, Ross, R. H., and Skole, D. L. (1993) Relating land use and global land cover change. IGDP report no. 24; HDP report no. 5.

Turner II, B. L., Skole, D. L., Sanderson, S., Fischer, G., Fresco, L. O., and Leemans, R. (1995) Land-Use and Land Cover Change - Science/Research Plan. IGBP Report No. 35; HDP Report No. 7. Stockholm and Geneva.

Turner M.D. (1999) Merging Local and Regional Analyses of Land-Use Change: The Case of Livestock in the Sahel. Annals of the Association of American Geographers 89, 191–219.

Turner M.G. and Gardner R.H. (1992) Quantitative Methods in Landscape Ecology: an introduction. In: Turner M.G., Gardner R.H. (eds). Quantitative methods in landscape ecology. Springer Verlag, New York.

van den Top G.M. (1998) The social dynamics of deforestation in the Sierra Madre, Philippines. Thesis Leiden University, Leiden.

van der Meer J., van Noordwijk M., Anderson J., Ong C. and Perfecto I. (1998) Global change and multi-species agroecosystems: concepts and issues. Agriculture, Ecosystems and Environment 67, 1–22.

Vanclay J.K. (1998) FLORES: for exploring land use options in forested landscapes. Agroforestry forum 9, 47–52.

Vayda A.P. (1983) Progressive contextualization: methods for research in human ecology. Human Ecology 11, 265–281.

Veldkamp A. and Fresco L.O. (1996) CLUE-CR: an integrated multi-scale model to simulate land use change scenarios in Costa Rica. Ecological Modelling 91, 231–248.

Veldkamp A. and Fresco L.O. (1997) Reconstructing land use drivers and their spatial scale dependence for Costa Rica. Agricultural Systems 55, 19–43.

Verburg P.H. and Chen Y.Q. (2000) Multi-scale characterization of land-use patterns in China. Ecosystems 3, 369–385.

Verburg P.H., Chen Y.Q. and Veldkamp A. (2000) Spatial explorations of land-use change and grain production in China. Agriculture, Ecosystems and Environment 82, 333–354.

Verburg P.H. and Denier van der Gon H.A.C. (2001) Spatial and temporal dynamics of methane emissions from agricultural sources in China. Global Change Biology 7, 31–47.

Verburg P.H., Veldkamp A., de Koning G.H.J., Kok K. and Bouma J. 1999a. A spatial explicit allocation procedure for modelling the pattern of land use change based upon actual land use. Ecological Modelling 116, 45–61.

Verburg P.H., Veldkamp A., Fresco L.O. (1999b) Simulation of changes in the spatial pattern of land use in China. Applied Geography 19, 211–233.

Walker R., Moran E.F., Anselin L. (2000) Deforestation and Cattle Ranching in the Brazilian Amazon: External Capital and Household Processes. World Development 28, 683–699.

Walsh S.J., Evans T.P., Welsh W.F., Entwisle B. and Rindfuss RR. (1999) Scale-Dependent Relationships between Population and Environment in Northeastern Thailand. Photogrammetric Engineering & Remote Sensing 65, 97–105.

Watson M.K. (1978) The scale problem in human geography. Geografiska Annaler 60B, 36–47.

Wear D.N., Turner M.G., Naiman R.J. (1998) Land cover along an urban-rural gradient: implications for water quality. Ecological Applications 8, 619–630.

Wilbanks T.J. and Kates R.W. (1999) Global changes in local places: How scale matters. Climatic Change 43, 601–628.

Wu F. (1998) Simulating urban encroachment on rural land with fuzzy-logic-controlled cellular automata in a geographical information system. Journal of Environmental Management 53, 293–308.

Zimmerer K.S. (1991) Wetland production and smallholder persistence: agricultural change in a highland Peruvian region. Annals of the Association of American Geographers 81, 443–463.

Chapter 3

GLOBAL WARMING AND THE ECONOMICS OF LAND-USE AND OF LAND-COVER CHANGE

R.A. Groeneveld
Environmental Economics and Natural Resources Group, Wageningen University and Research Centre

G. Kruseman
Development Economics Group, Wageningen University and Research Centre

E.C. van Ierland
Environmental Economics and Natural Resources Group, Wageningen University and Research Centre

1.   Introduction

Land-use and land-cover change (LUCC) is driven by a combination of:

(1) biophysical factors which determine the capability of land use;

(2) technological and economic considerations which determine the socio-economic feasibility of land use and

(3) institutional and political arrangements which determine the acceptability of land use (e.g., Barlowe, 1986; Turner *et al.*, 1995).

The complexity of land-use and land-cover change is also manifested in the multifunctionality of land: climate change policies interact with several other policy issues such as food production, forestry, recreation and biodiversity conservation.

In this chapter we discuss some important economic topics on land-use and land-cover change, climate change, and policy options that stem from economic research on climate change mitigation. We wish to present the economic issues of land use to researchers and policy makers involved in climate change policies. First, we discuss theories underlying economic explanations of LUCC and climate change. Here the focal point is the economic choice problem of

A.J. Dolman et al. (eds.), Global Environmental Change and Land Use, 53-69.

the land user who seeks to maximize a certain goal within given biophysical, technological and institutional constraints. Economic theory has also provided LUCC research with economic models of land-use and land-cover change (see also Chapter 2). Second, we discuss policy recommendations for mitigation of climate change through land use policy within the economic analytical framework that economists have offered for environmental policy and spatial planning. Finally, to illustrate the role of important economic variables in LUCC, we discuss two case studies illustrating the relation between LUCC and climate change. The first case study deals with CH_4 emissions from rice paddies. The second case study refers to reducing net greenhouse gas emissions by forestation, afforestation, fallow management and pasture management in Mexico. The chapter concludes with a discussion on climate policy options with regard to LUCC. Although the scope for CO_2 emission reduction and carbon sequestration through land use policy is limited, the available options are in many cases cost-effective, provided that transaction costs can be kept within reasonable limits. The contribution of land use policy to CH_4 and N_2O emission reduction can be more substantial because agricultureagriculture is one of the main sources and several low-cost options are available to reduce CH_4 and N_2O emissions from agricultureagriculture.

2. Economic theoretical analyses of LUCC and climate change

Economists analyze problems of choice for a decision-maker, often in a setting with limited resources and unlimited wants. By finding rational solutions for choice problems and rational explanations for observed choices, economics analyzes the role of economic incentives in human behavior, and offers recommendations for public policy.

This section gives a brief outline of the economic approach to the analysis of LUCC and its relation to climate change. We will first discuss the main elements in economic analyses of LUCC before we focus on the specific relation of LUCC to climate change. The section concludes with a discussion on the role of information from the natural sciences in economic models of LUCC.

2.1 The economic analysis framework of LUCC decisions

Parallel to Hazell and Norton's (1986) characterization of agricultural sector models, economic analysis of LUCC generally includes five elements:

(1) the land user's economic behavior and goals;

(2) the currently and potentially available production technology sets;

(3) the resource endowments and their distribution;

(4) the markets for products and inputs needed for production;

(5) the policy environment.

Economic behavior and goals of decision making agents The decision making agents in economic analyses of LUCC are land users such as individuals, enterprises, government agencies or NGOs, with varying objectives such as profit maximization, long-term subsistence or nature conservation. These agents pursue their objectives by means of the limited resources available within a given institutional and informational framework.

The majority of economic models take an optimization approach to economic decisions that maximizes an objective variable within given constraints. Experimental economists and behavioral economists (e.g., Davis and Holt, 1993) have developed alternative models of economic behavior based on insights from experiments and psychology. Jager *et al.* (1999) include psychological models of economic behavior in a multi-agent model to perform analyses on a society scale.

Production technology sets Besides the objectives of the economic agent, some specification exists either implicitly or explicitly of the relations between inputs and outputs. The specification of these relations in production functions thereby constitutes an important bridge between economic and biophysical sciences (Kruseman and van Keulen, 2001). Biophysical process models are combined with economic decision making to determine allocative efficient solutions from both an economic and an environmental point of view. One should note, however, that the concepts of efficiency and optimality are used in different ways by economists and biophysical scientists.

Resource endowments and the role of space and location The central resource in any analysis of LUCC is land, which has unique characteristics: a fixed location, a given geoclimatic environment, a relatively limited supply, and in many cases a strong heterogeneity in quality (Miranowski and Cochran, 1993). Quality, in economic terms, depends on a number of local circumstances, such as distance to factor and product markets, and suitability for crop production. Land (or rather location) quality is also related to other resource endowments, such as available labor force (population density) and fresh water. Location and the correlating quality of the land resource play therefore an important role in the economic choice problem of the land user. Economic analyses of LUCC generally investigate where activities are allocated, or which plots will be devoted to which land-use and land cover type. Although these two questions essentially deal with the same problem, they characterize the two branches of economics involved in the analysis of LUCC, namely regional economics and natural resource economics.

Regional economics (e.g. Isard *et al.*, 1998; Fujita *et al.*, 1999) investigates the role of the spatial dimension in economic activities, including changes in economic activity that affect spatial land use structures. Much research in regional economics focuses on external economies of scale of industrial activities and is therefore often classified as urban economics (Cheshire and Mills, 1999). The classical Von Thünen model (see Wartenberg, 1966), which deals with spatial patterns of agricultural land use as a function of the distance from a single market, is one of the first analyses of spatial land use decisions and still serves as a classical example of spatial economic analysis.

Studies in natural resource economics investigate the optimal use of natural resources. The branch of natural resource economics that focuses on land as a natural resource is referred to as land economics (Barlowe, 1986; van Kooten, 1993). Land economics deals explicitly with the economic relationships among people concerning land. Although land is used in many ways (including urban, recreational and mining uses), land economics has often focused on agricultural land uses, as agricultureagriculture accounts for most of earth's surface with economic value (Barlowe, 1986).

Markets   Markets determine the allocation of inputs and outputs based on the self-interest of economic agents. The supply of commodities and demand for inputs is shaped by the relevant production functions. The demand for food and fiber determined by population densities, income levels and relative prices of alternatives shape the demand side of markets. Distance, transportation costs and quality differences of products and endowments determine price differentials within the marketing chain. Comparative advantages and economies of scale for certain types of land use based on market considerations and agro-ecological circumstances help to determine the spatial and temporal distribution of different land uses in the global landscape.

The land market also influences LUCC, since the price of land influences the relative profitability of certain productive activities. The land price is determined by the opportunity costs of alternative land allocation. In densely populated industrialized areas, land prices tend to be higher than in sparsely populated rural areas, since in the former there are many claims on small pieces of land for highly productive and highly valued consumptive activities (in terms of value per unit of land) relative to the latter.

The policy environment   Government policy affects the economy through instruments such as taxes, subsidies, quotas and legal restrictions. In the case of land use, major impacts occur through administered agricultural prices or the provision of subsidies. Also policies for nature conservation are becoming more important. In forestry, the assignment of property rights and royalty systems are extremely important incentives for either deforestation or conservation of forests. Tradable quotas for greenhouse gas emissions, or taxes on the emissions of greenhouse gases will also have a direct impact on land use decisions. In most cases, private decisions will determine how land is allocated, on the basis of incentives provided through the market or by governments.

2.2   Economic analysis of LUCC and climate change

Within the economic analysis framework described in the previous section it is possible to identify problems in the relation between LUCC and the global climate. The typical questions asked in this type of analysis are:

(1) What is the optimal allocation of resources, given their impact on the environment?

(2) Why do economic decisions of individuals sometimes lead to socially non-optimal allocations?

(3) What can be done to achieve a socially optimal allocation?

Economic explanations of environmental pollution and recommendations for environmental policy are mainly based on theories of market or government failure. Theoretically, efficient markets provide resource allocations that are both individually and socially optimal: therefore, resource allocations (i.e., land use patterns) with excessive trace gas emissions indicate that existing markets are not efficient: they fail in achieving a social optimum. The main sources of market failure in the economics of land are:

(1) externalities;

(2) ill-defined property rights that underlie the public good nature of environmental goods; or

(3) divergence of time preferences (Miranowski and Cochran, 1993).

Externalities   Externalities play a major role in economic analyses of environmental problems (van Kooten, 1993). In the definition of Baumol and Oates (1988), an externality

> is present whenever some individual's (say A's) utility or production relationships include real (that is, non-monetary) variables, whose values are chosen by others (persons, corporations, governments) without particular attention to the effects on A's welfare. The decision maker, whose activity affects other's utility levels or enters their production functions, does not receive (or pay) in compensation for this activity an amount equal in value to the resulting benefits (or costs) to others.

For climate change this means that the well-being of many is affected by, say, the trace-gas emissions of some, who do not pay for the damage they inflict on others. On the other hand, carbon sequestration is not likely to be undertaken by individual land owners and land users as long as there are no net benefits to them individually, even though society as a whole will profit from it. This discrepancy between private and social costs leads to suboptimal land use decisions (Nordhaus, 1991; van Kooten, 1993). Providing the proper incentives to individual economic agents should therefore be an essential element in climate change policy. The implications of externalities for climate change and land use policies are in practice very complex and related to the definition of property rights and the issue of discounting and time preference.

Property rights   According to standard economic theory, market economies channel the self-interest of individuals such that an allocation is found that is desirable for all parties, as long as property rights are well defined. A good definition of property rights requires that they are:

(1) completely specified, such that ownership is clearly delineated and that restrictions of ownership and penalties for violation of property rights are specified;

(2) exclusive, i.e., the owner determines who may use the property and under what conditions;

(3) transferable;

(4) enforced (van Kooten, 1993). Much environmental pollution and resource overuse can be explained from failure to meet these requirements.

Whereas much land is privately owned, the atmosphere, or so to say, 'the climate' has no owner and therefore no means exist to exclude anyone from using it: so far there are no property rights to transfer or enforce. This public good nature of the climate provides no incentives to individuals to improve its quality, because the individual benefits of such action are negligible while the costs are substantial. The protection and maintenance of public goods is generally an important justification for government intervention, and therefore the public good nature of climate change is the main reason why government intervention and international co-operation is generally recommended to mitigate it. The system of tradable emissions permits in climate change policies is aiming at establishing clear entitlements to the right to emit greenhouse gases. In practice this leads to complicated questions of how to distribute these rights and which principles for grandfathering or auctioning should be used. Whether emissions of greenhouse gases that occurred in the past should also be taken into account is an interesting and highly debated issue.

Time preference   Time preference is a common factor in economic decisions reflecting interest rates as well as a general sense of impatience. There are indications that the social discount rate, i.e., the rate of time preference apparent in public decisions, differs from the private discount rate, which is the rate of time preference apparent in decisions by individuals. The main reason for this divergence is that, in contrast to individuals, society is supposed to take the demands of future generations into account: a high discount rate renders future interests (and hence those of future generations) less important than a low discount rate. Related to this argument is the observation that individuals have different discount rates as consumers than as members of a society: in a political context citizens may express a desire for a lower discount rate than is usually revealed by individual saving decisions (Hanley and Spash, 1993). These arguments suggest that divergence between the private and social discount rates implies that the social discount rate is generally lower than the private discount rate. Hence, decisions on public goods are not likely optimal when based on a private discount rate.

In the climate debate the question is raised whether discounting of future costs and benefits is justified if long term irreversible damages are at stake. Discounting of future damages may in the end lead to irreversible changes that policymakers would never have chosen if they had been aware of their full and irreversible impacts.

The role of biophysical and institutional factors in economic analyses

Biophysical and institutional factors are important in any analysis of LUCC. Agricultural land uses, which have a high share in land-cover area and an important role to play in climate change mitigation, depend strongly on biophysical processes and variables. Furthermore, spatial planning policy, urbanisation and nature conservation restrict the area available to farmers and foresters. Therefore, the economic choice problem of the land user takes place within biophysical and institutional constraints (Figure 3.1).

There exists some flexibility in these constraints. For example, spatial planning policy can be influenced by land users, in particular when the economic stakes are high. Land users can also improve the suitability of their land by fertilization and irrigation, although such institutional and biophysical adaptations can only be made at considerable costs. Therefore, institutional and biophysical factors, as well as their spatial dimension, are considered explicitly in economic models of LUCC. Many of these models typically include biophysical constraints as spatially varying land suitability characteristics. Linear and non-linear programming models treat land suitability, depending on biophysi-

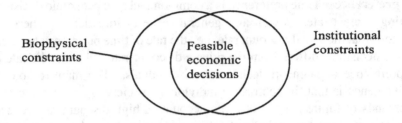

Figure 3.1. The economic choice problem of a land user takes place within biophysical and institutional constraints.

cal as well as geographic and demographic considerations, as a characteristic of locations that determines the comparative advantage of plots or regions.

Apart from traditional agricultural economic models, an economic modeling approach referred to as bio-economic modeling has evolved in response to the important role of biophysical processes in agricultureagriculture and forestry. The term bio-economic modeling is used to stress the importance of interdisciplinarity in the approach. Kruseman (2000) defines bio-economic modeling as

> a quantitative methodology that adequately accounts for biophysical and socio-economic processes and combines knowledge in such a way that results are relevant to both social and biophysical sciences.

The key issues in this definition refer to the synergy between biophysical and socio-economic sciences.

The demand for data from different sources and disciplines and on different scale levels remains a bottleneck for interdisciplinary research. To apply spatially explicit models with a satisfactory grain and inputs from several sources, tools are needed to connect the models to the available data and to ensure that models and data apply to more or less the same scale level. The modeler faces the choice between either adapting the model to the data, or to apply econometric techniques that can fill in the gaps (Groeneveld and van Ierland, 2000). Although many economic models have a high level of detail which often does not suit the availability of spatial data, spatially explicit models have been developed that can be estimated with limited data (e.g., Oude Lansink and Oskam, 1995; Chomitz and Gray, 1996; Geoghegan *et al.*, 1997). On the other hand, statistical tools such as Maximum Entropy Estimation (Golan *et al.*, 1996) have been designed to extract as much information from the data as possible. An application of Maximum Entropy Estimation to land use models can be found in Miller and Plantinga (1999) and Groeneveld and van Ierland (2000).

3. The policy context

In the discussion on land use policies for mitigating climate change, the emphasis is often put on direct measures in terms of emission quotas, taxes on greenhouse gas emissions or subsidies on carbon sequestration. However, before we discuss polices targeted directly at climate change mitigation, it is useful to sketch the general context of LUCC, as the potential impact of policies that are directed at agricultureagriculture and land use (and not directly related to climate change policy) is tremendous. We focus hereby on Europe.

From its introduction in the early 1960s, the European Union's Common Agricultural Policy (CAP) was first directed at production increases and efficiency improvements. This has resulted in an intensification of agricultureagriculture and livestock. Burning of crop residues, common in many parts of Europe in the 1960s has all but disappeared today, contributing to carbon dioxide emissions. Livestock numbers have increased tremendously, in particular in relation to the pasture area (much higher apparent stocking rates). This is partly due to more intensive use of pastures and to livestock systems where animals no longer make use of pastures for grazing. Intensification of agricultureagriculture has also led to increased energy use and fertilizer application in agricultural land use activities, which also contributes to greenhouse gas emissions. Policy directed at the control of air pollution related to acid rain through control of animal waste also has side-effects on the emission of greenhouse gasses.

The present state of European agricultureagriculture is characterized by over production in some sub-sectors. This has led to a policy of set-aside arrangements to take land out of production for a short fallow period. This policy can be combined with climate change policy to bring incentives for land use that mitigates the emission of greenhouse gasses.

Probably, the introduction of milk quotas has been one of the most important factors in LUCC in the EU. The dairy herd has decreased by more than 25% between 1985 and 1996 in the Netherlands. The effect of the reduction was offset partly by an increase in sheep and goat numbers (Brouwer and van Berkum, 1998).

Agriculture does not produce only food and fibre, as is apparent from the original objectives of the CAP, but also shapes the rural landscape, as becomes apparent in structural changes of CAP reform. The main considerations in the agricultureagriculture-environment linkages in CAP reform have been geared towards curbing nutrient surpluses, especially in areas with intensive livestock systems and decreasing pesticide use. The original CAP with its emphasis on production and price support encouraged high levels of chemical inputs (Brouwer and Lowe, 1998).

3.1   Policies to mitigate climate change through land use

An important challenge for climate policies is to provide proper incentives for land-use and land-cover changes that contribute to cost-effective carbon sequestration or a net reduction of greenhouse gas emissions. These policies focus on:

- Biomass energy (e.g., short rotation willow, switch grass or oil seeds);

- Reforestation and afforestation;

- Reduction of deforestation (particularly of high quality tropical rain-forests and forest with high biodiversity values);

- Improvement of soil organic matter (which also helps in reducing soil erosion and boosting agricultural production);

- Reduction of N_2O losses through mineral policies;

- Reduction of CH_4 emissions from rice paddies (e.g., through water man-agement or use of fertilizer) or natural ecosystems (e.g., bogs and fens).

Given the rapid expansion of the economy and population growth in the world, strong pressure exists on land as a resource. Competition for food production, nature conservation, carbon sequestration, energy supply (biomass, wind power, solar power) and other infrastructure occurs in many countries in the world, and will be even more severe in the future.

Biomass contributes at present about 9–13% of the world's energy supply. The potential role of biomass in future energy supply is large according to various studies quoted in Turkenburg (2000) and total supply might go up from about 45 exajoules at present to about 200 exajoules by the year 2050. In planning these activities, spatial policies should provide proper incentives for efficient land use, multifunctional use of land and for net reduction of greenhouse gas emissions where this can be cost-effectively achieved. The competitiveness of biomass will depend on the prices of fossil fuels, but also on the costs and net returns from alternative, competing uses of productive land (Turkenburg, 2000). Multiple use of land, e.g., combining biomass production or carbon sequestration with recreation, may reduce greenhouse gas concentrations at minimum costs. Whether biomass production or carbon sequestration can compete with, or is complementary to, other forms of land use is an important question determining the amount of land available for net greenhouse gas emission reductions (Watson *et al.*, 2000, see also Chapter 8).

Tradable emission permits, taxes on greenhouse gases, Joint Implementation (JI) and the Clean Development Mechanism (CDM) are economic instruments that will contribute to changes in land use and resulting emissions of greenhouse gases. Tradable emission permits provide incentives to reduce GHG emissions, because the price of a permit provides an incentive to reduce emissions at the lowest costs. The same holds for emission taxes. Under the Kyoto protocol JI and the CDM are envisaged.

For using land use in mitigating climate change these mechanisms are attractive, because land is widely available in Eastern Europe and in developing countries. In these areas low cost opportunities for biomass production or carbon sequestration are available that can be exploited under JI or under the CDM. Several problems occur in the application of these instruments:

(1) Can it be guaranteed that the project to be financed is additional to existing projects, i.e., would the project not be carried out without additional financial support?

(2) Can proper measurement and monitoring take place, given the dynamics of the natural systems (e.g., precipitation, temperature fluctuations and diseases) and political instability?

(3) How can it be avoided that carbon release takes place at other locations than the project areas (carbon leakage)?

To what extent land-use changes can be expected in the future largely depends on future climate policies. Strong incentives will be created for the production of biomass, for carbon sequestration and other options to reduce the net greenhouse gas emissions through changes in land-use and land-cover when relatively high emission reduction costs for greenhouse gases are necessary. Moreover, a balance will be established with competing land use activities, through changes in the price of land and the prices of the goods and services produced.

4. Case studies of LUCC and global warming

The previous sections gave an outline of the economic driving forces of LUCC and their relation to the global climate. In this section we will describe two case studies of land use and climate policy to illustrate how economic and biophysical variables interact in practice. Land use decisions are made by households or farms to undertake productive activities using their scarce resources (labor, land and capital) to attain their goals and aspirations. The economic circumstances facing households and farms are the key driving force of land-use change.

4.1 Technological development in Asian rice systems

Agriculture is responsible for two thirds of anthropogenic emissions of atmospheric CH_4. Within the agricultural sources of anthropogenic CH_4, the emissions from rice paddies figure prominently. Measurements made in flooded rice indicate that CH_4 fluxes in flooded rice depend critically on climate, soils, paddy type and agricultural practices (Mosier *et al.*, 1998). A 15–20% decrease in CH_4 emissions is necessary to stabilize the atmospheric concentration at the present levels (Watson *et al.*, 2000).

Under influence of the Green Revolution and population growth in Asia both production and productivity of rice have gone up over the past decades, although both growth trends tend to diminish at present. Understanding the driving forces behind the changes that took place in agricultural development in Asia as well as the biophysical processes involved in CH_4 emissions allows us to understand the effect of this type of land-use change on global warming.

Green Revolution technologies in Asia are generally based on the use of irrigation schemes in combination with intensive use of fertilizers, and high yielding rice varieties. The increase of flooded rice area has led to increased CH_4 emissions. The green revolution was possible because economic incentives (irrigation schemes, credit, input subsidies) were in place, making it profitable for some farmers to adopt the technology packages of improved seed, fertilizer and biocides.

Table 3.1.   Estimates of methane emissions from rice fields in 1994. Source: Mosier (1998).

Country	Total area of rice paddies (10^{10} m^2)	Total rice grain yield (Mg)	CH_4 emission (Tg y^{-1})	
China	32.2	174.7	13.0	— 17.0
India	42.2	92.4	2.4	— 6.0
Japan	2.3	13.4	0.02	— 1.04
Thailand	9.8	19.2	0.5	— 8.8
Philippines	3.5	8.9	0.3	— 0.7
World total	147.5	473.5	25.4	— 54.0

Table 3.1 shows that the main rice producers account for 63% of rice area and methane emissions. Especially in China the emissions are relatively high. The main differences can be attributed to nutrient management. Especially the application of rice straw greatly enhances methane emissions, while sulfate and nitrate fertilization decreases the emissions.

The benefits of the green revolution have accrued firstly to large farmers in Asia. Only after new high yielding varieties were well in place, did small or isolated farmers start to benefit from these, albeit with less profit than large farmers (Lipton and Longhurst, 1989). The important issue for climate change is that there are large differences in the way large and small farmers apply external inputs. Poor farmers generally rely more on fertilization with straw, manure and compost, while large farmers use recommended technology packages. This difference also is apparent in more densely and sparsely populated areas in China (Bray, 1986). Organic and low external input agricultureagriculture in flooded rice systems tends to be much more methane emitting than mainstream high external input systems.

Economic liberalization in China in the past decade has lead to a situation where in many rice producing areas there is stiff competition for labor. As a result the traditional use of green manure has increasingly been replaced by the use of inorganic fertilizers. Although this appears to have a positive effect on the emission of methane, there are increasing problems with compacted soils due to mechanized land preparation with tractors. There is a clear trade-off between soil quality and greenhouse gas emissions, in addition to a trade-off between a reduction methane emissions and economic growth.

The problem of methane emissions is linked to the flooding of paddy fields. There is no direct crop physiological reason why these fields are flooded, rice produces well under reduced water regimes. It is probable that the choice of flooding rice was made centuries ago to lower weeding requirements in an era when water was abundant relative to labor at weeding time. Research will have to show if low water rice is a viable alternative under the present conditions where water is becoming more scarce, with the additional benefit of reducing methane emissions.

4.2   Agroforestry in Mexico

This case study on forestry and agroforestry measures in the Central Highlands of Chiapas in Mexico is based on de Jong (2000). He analyzes the costs of reducing greenhouse gas emissions by forestation, fallow management and agricultural and pasture management in Mexico. The study takes into account the private costs of these management options and the opportunity costs of agricultural income foregone.

The study pays extensive attention to the risk of carbon leakage, i.e., the option that additional deforestation will take place at another location, outside the project boundaries. If this type of carbon leakage occurs, a project-based approach should also pay attention to these problems, in order to avoid severe overestimation of the emission reductions.

The project shows that reforestation can result in relatively large reductions of CO_2 emissions at relatively low costs of 5–20US$ per ton of carbon. Within this cost range, forestry and agroforestry measures in the relevant study area could mitigate from 1 to 42 tons of carbon, with a maximum economic supply of carbon sequestration of around 55 ton carbon at 40 US$ per ton of carbon. If it could be granted that indeed the forestation project would not be carried out without financial support from Western countries (i.e. that the project is additional to existing projects) and that no carbon leakage would take place, the project would be useful for the CDM. In this case, however, it is not clear whether reforestation would not have occurred if no climate policies were implemented. It is therefore emphasized that it is difficult – if not impossible – to identify an unbiased scientific method to make a judgement about the additionality of a project. Specific attention should also be given to the transaction costs of these land use policies for mitigation of climate change. If a large number of smallholder farmers participate in the project, considerable transaction costs are involved and the cost of mitigation through CDM will rapidly increase.

Whether land use and land management will indeed change as a result of climate policy depends on the economic incentives that will be provided and the institutional setting under which this will take place. If financial remuneration will be given for management leading to carbon sequestration, and if the management system is consistently applied and well monitored, then a substantial amount of carbon sequestration may be achieved. By introducing economic incentives for carbon sequestration some of the externalities of greenhouse gas emissions will be internalized.

5.   Discussion and conclusions

Although changes in land-use and land-cover can contribute only modestly to reduction of atmospheric CO_2 concentrations, some of the available options (e.g., carbon sequestration through afforestation) are cost-effective, in particular in countries with low prices for land and food. Changes in market and price policies can sometimes (as in the case of milk quotas in the EU), change the intensity of land use. Total emissions of methane from livestock dropped 6% in the EU between 1985 and 1996 (Hutjes *et al.*, 2001). It is not probable that emissions under business as usual will drop further since it is the result of a shock policy (the introduction of a system of milk quotas) that is not likely to be replicated. In addition there is substantial scope for the reduction of other trace gas emissions such as CH_4, for example in rice agricultureagriculture, or the management of wetlands. Under the Kyoto protocol, JI and the CDM offer scope for financing land-use and land-cover projects to reduce the net emissions of greenhouse gases. Developing countries would be able to protect valuable ecosystems, financed by industrialized countries, whereas the latter will be able to reduce net greenhouse gas emissions at low costs.

If land-use and land-cover change is to fulfil its potential contribution to trace gas concentration reductions, individual land owners and land users should face the proper incentives in their decisions. Because of the public good nature of the global climate, many positive and negative effects of certain land use types are not incorporated in the prices on which decisions are based. For example, taxation of CO_2-emissions from fossil fuel and (partial) duty exemptions for

CO_2-neutral biofuels reflect the social costs and benefits associated with these commodities, which may stimulate a shift in favor of biofuels.

It is, however, essential that the policies are implemented such that they provide no incentives to destroy virgin forest or to convert them into biomass plantations with low values for nature conservation and biodiversity protection. The multi-functionality of land implies that climate policy may conflict with other objectives, although synergies are also possible. Policies for carbon sequestration could be integrated with spatial planning, agricultural policies and policies for sustainable energy systems. If climate policies are well embedded in existing policy areas, and if economic instruments and flexible mechanisms are implemented, considerable scope exists for multi-functional land, where carbon sequestration, production of sustainable energy and sustainable agricultureagriculture are – at least partially – combined.

References

Barlowe, R. (1986) Land resource economics: the economics of real estate. Prentice Hall, Englewood Cliffs.

Baumol, W.J. and Oates, W.E. (1988) The theory of environmental policy. Cambridge University Press, Cambridge.

Bray, F. (1986) The rice economies: technology and development in Asian societies. Basil Blackwell, London.

Brouwer, F. and Lowe, P. (1998) CAP reform and the environment. In: Brouwer, F. and Lowe, P. (eds) CAP and the rural environment in transition. Wageningen Pers, Wageningen, 13–38.

Brouwer, F. and van Berkum, S. (1998) The Netherlands. In: Brouwer, F. and Lowe, P. (eds) CAP and the rural environment in transition. Wageningen Pers, Wageningen, the Netherlands, 167–184.

Cheshire, P. and Mills, E.S. (eds) (1999) Handbook of regional and urban economics: applied urban economics. Amsterdam Publishers, Amsterdam.

Chomitz, K.M. and Gray, D.A. (1996) Roads, land use, and deforestation: a spatial model applied to Belize. World Bank Economic Review 10, 487–512.

Davis, D.D. and Holt, C.A. (1993) Experimental economics. Princeton University Press, Princeton, New Jersey.

de Jong, B.H.J. (2000) Forestry for mitigating the greenhouse effect: an ecological and economic assessment of the potential of land use to mitigate CO_2 emissions in the highlands of Chiapas, Mexico. PhD Thesis, Wageningen University, Wageningen, the Netherlands.

Fujita, M., Krugman, P. and Venables, A.J. (1999) The spatial economy: cities, regions and international trade. The MIT Press, Cambridge, Massachusetts.

Geoghegan, J., Wainger, L.A. and Bockstael, N. (1997) Spatial landscape in-
dices in a hedonic framework: an ecological economics analysis using GIS.
Ecological Economics 23, 251–264.

Golan, A., Judge, G. and Miller, D. (1996) Maximum entropy econometrics:
robust estimation with limited data. John Wiley & Sons, Chichester.

Groeneveld, R.A. and van Ierland, E.C. (2000) Economic modelling approaches
to land use and cover change. National Research Programme on Global Air
Pollution and Climate Change, Bilthoven, The Netherlands.

Hanley, N. and Spash, C.L. (1993) Cost-benefit analysis and the environment.
Edward Elgar Publishing Ltd., Aldershot, United Kingdom.

Hazell, P.B.R. and Norton, R.D. (1986) Mathematical programming for economic analysis in agriculture. MacMillan Publishing Company, New York.

Hutjes, R.W.A., Huygen, J., Nabuurs, G.J., Schelhaas, M.J., Verhagen, A. and
Vleeshouwers, L.M. (2001) Stocks and fluxes of carbon and other land use
related greenhouse gas emissions. In: Dolman, A.J. (ed.) Land Use, climate
and biogeochemical cycles: feedbacks and options for emission reduction.
NRP, Bilthoven, the Netherlands.

Isard, W., Azis, I.J., Drennan, M.P., Miller, R.E., Saltzman, S. and Thorbecke, E.
(1998) Methods of interregional and regional analysis. Ashgate, Aldershot,
UK.

Jager, W., Janssen, M.A. and Vlek, C.A.J. (1999) Consumats in a commons
dilemma: testing the behavioural rules of simulated consumers. Centre for
Environmental and Traffic Psychology, Groningen.

Kruseman, G. (2000) Bio-economic household modelling for agricultural intensification. PhD Thesis, Wageningen University, Wageningen, the Netherlands.

Kruseman, G. and van Keulen, H. (2001) Soil degradation and agricultural production: economic and biophysical approaches. In: Heerink, N., van Keulen,
H. and Kuiper, M. (eds) Economic policy and sustainable land use: Recent
advances in quantitative analysis for developing countries. Physica Verlag,
Heidelberg, 21–48.

Lipton, M. and Longhurst, R. (1989) New seeds and poor people. Unwin Hy-
man, London.

Miller, D.J. and Plantinga, A.J. (1999) Modeling land use decisions with aggregate data. American Journal of Agricultural Economics 81, 180–194.

Miranowski, J.A. and Cochran, M. (1993) Economics of land in agriculture.
In: Carlson, G.A., Zilberman, D. and Miranowski, J. A. (eds) Agricultural
and environmental resource economics. Oxford University Press, New York,
392–440.

Mosier, A.R., Duxbury, J.M., Freney, J.R., Heinemeijer, O., Minami, K. and
Johnson, D.E. (1998) Mitigating agricultural emissions of methane. Climatic
Change 40, 39–80.

Nordhaus, W.D. (1991) The cost of slowing climate change: a survey. The Energy Journal 12, 37–66.

Oude Lansink, A. and Oskam, A. (1995) Land-share analysis of EU crop production. Tijdschrift voor Sociaal wetenschappelijk onderzoek van de Landbouw 10, 174–190.

Turkenburg, W. (2000) Renewable energy technologies. UNDP, New York.

Turner, B.L., Skole, D., Sanderson, S., Fisher, G., Fresco, L. and Leemans, R. (1995) Land-Use and Land-Cover Change, Science/Research Plan. IGBP, Stockholm.

van Kooten, G.C. (1993) Land resource economics and sustainable development: economic policies and the common good. UBC Press, Vancouver.

Wartenberg, C.M. (1966) Von Thünen's Isolated State. Pergamon Press, Oxford.

Watson, R.T., Noble, I.R., Bolin, B., Ravindranath, N.H., Verardo, D.J. and Dokken, D.J. (eds) (2000) Land use, land-use change, and forestry. Cambridge University Press, Cambridge, UK.

III

THE RELATION BETWEEN CLIMATE
AND LAND USE

Chapter 4

LAND COVER AND THE CLIMATE SYSTEM

R.W.A. Hutjes
Alterra, Wageningen University and Research Centre

P. Kabat
Alterra, Wageningen University and Research Centre

A.J. Dolman
Vrije Universiteit Amsterdam

1. Introduction

The earth's land-cover and vegetation interact with the atmosphere by several inter-linked mechanisms (Figure 4.1). Vegetation controls evaporation through its internal physiology: the opening of stomata responds to environmental conditions such as temperature, humidity deficit, radiation, CO_2 concentration, and soil moisture. The opening of the stomata also affects the rate of photosynthesis. Stomata respond rather quickly to changes in environmental conditions. A characteristic time scale for coupling with the atmosphere is typically 0.01–1 hrs (strong coupling). More passively, through radiative properties and aerodynamic characteristics, vegetation affects partitioning of the surface energy balance and also dynamics of the wind. Although the coupling between these surface characteristics (e.g., albedo, roughness) is determined by very fast, turbulent processes (typical time scales in seconds), a significant change in albedo would normally follow the seasonal evolution of vegetation with time scales of a week to a year (moderate coupling).

A.J. Dolman et al. (eds.), Global Environmental Change and Land Use, 73-110.
© 2003 *Kluwer Academic Publishers.*

Importantly, vegetation also moderates the biogeochemical and hydrological cycles by changing the pools and fluxes of nutrients and sediments, and by interfering with the lateral redistribution of water and transported constituents. Vegetation and soils exchange large quantities of carbon dioxide and other greenhouse gases with the atmosphere, thereby influencing the overall radiation balance of the earth. Coupling of biogeochemical and soil chemical processes with the atmosphere is most significant at very long time scales: one to several hundreds of years (weak coupling). But, arguably, in the context of 'climatic and global change' this coupling mechanism may appear to be one of the most crucial components of the system (Cox *et al.*, 2000, Friedlichstein *et al.*, 2001, see also Chapters 5 and 6).

Thus, the terrestrial biosphere, the atmosphere and the hydrological and bio-geochemical cycles are intrinsically coupled. The coupling mechanism in the biosphere-atmosphere-hydrosphere system is often non-linear, and generally bi- or multi-directional. This means that an individual component of the system is both an actor and receptor; is under the influence of, as well as impacting upon, the remaining parts of the system. The strength of the coupling varies with the characteristic time scales, which range from seconds to millennia. This coupling also implies that human modification of land-cover, conversion from natural to agricultural ecosystems and subsequent management practices, will have effects on other parts of the system. Local land-cover or land-use changes will have immediate effects on local hydrology, large scale changes may eventually affect global scale climate. In this particular section we will limit ourselves mostly to a discussion of biosphere-atmosphere-hydrosphere interactions in terms of energy and water fluxes. Biogeochemical interactions are discussed more extensively in Chapters 5 and 6 of this book.

Figure 4.2 schematically depicts the various feedbacks between land surface and atmosphere for three typical land surface characteristics: albedo, soil moisture and the aerodynamic roughness length. In each loop, positive and negative feedbacks occur, which implies that the overall effect on for example the rainfall will be a trade-off between these feedbacks. For instance, an increase in albedo, resulting from a land-cover change, will lead to a decrease of radiation absorbed by the land surface, which in turn may cause a decrease in latent and sensible heat fluxes to the atmosphere. This reduced heat and moisture input to the atmosphere could result in decreased cloudiness, less convergence and eventually in lower rainfall. Lower rainfall results in a decrease in surface soil moisture and, when persistent, in less vigorous vegetation, which both contribute to a further increase of the surface albedo (= positive feedback). However, less

Biosphere / Atmosphere Coupling

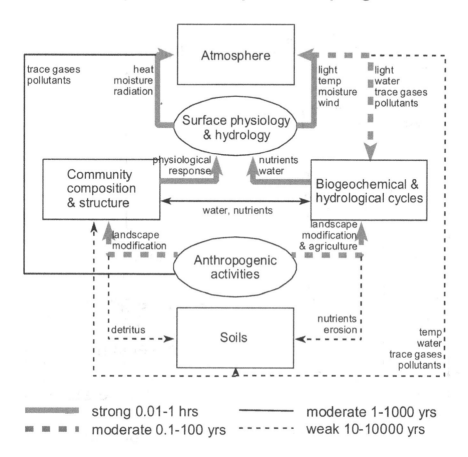

Figure 4.1. Ecological, hydrological and biogeochemical interactions between land surface and atmosphere.

cloudy conditions will lead to an increase of incoming radiation, and due to more available radiative energy at the surface, may lead to an increase of latent and sensible heat fluxes (= negative feedback). This subtle balance between mechanisms acting in opposing directions may introduce additional levels of non-linearity, and may increase uncertainties when making predictions.

Over the past twenty years a strong body of evidence of the effects of land surface modification on climate has emerged (Pielke *et al.*, 1998). Most of these studies were of the 'force - response' type, where the response of climate to prescribed changes in land-cover or to modifications of the land surface parameters, was analysed. More recently, two-way interactions between vegetation and climate have been brought into models.

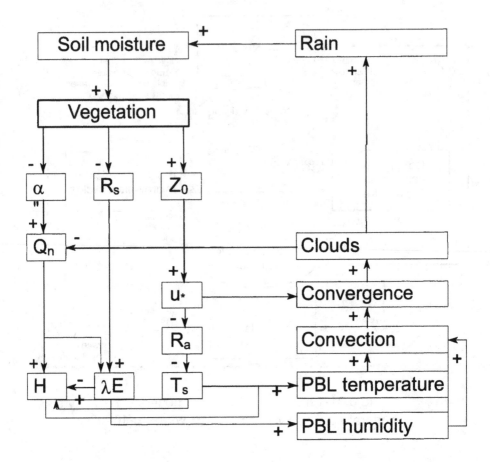

Figure 4.2.　Schematic representation of various interactions between land surface and atmosphere. Three basic interaction mechanisms are depicted: a) through the radiation balance; b) through energy partitioning and c) through turbulence and wind convergence. Symbols: α albedo, Q_n net radiation, H sensible and λE latent heat fluxes, Z_0 aerodynamic roughness length, u^* friction velocity, R_a aerodynamic resistance, R_s surface resistance, T_s surface temperature. See text for further explanation.

Much of this work focused on the effects of large-scale human-induced land-cover changes in the tropics. Desertification in semi-arid regions has been the subject of intensive research and debate since the landmark paper by Charney (1975), demonstrating that albedo effects, related to degradation of vegetation, may have played a role in the persistent drought of the last three decades in the Sahel. Our knowledge of that system has greatly increased since, as we will discuss below. Similarly, many studies have been conducted on the potential impact of deforestation in the wet tropics. One of the first studies for the Amazon (Henderson-Sellers and Hornitz, 1984) predicted a 15–20 mm per month decrease in rainfall following the complete removal of tropical rain forest and its replacement by pasture. Albeit more limited than for the Sahel, also in this case much has been learned since then.

There are two reasons that explain why much of this work focuses on the tropics. First, land-use changes presently occur mostly, and at unprecedented rates, in the tropics, though in the past changes of similar magnitude occurred in other parts of the globe (Mediterranean, Central US, East and South Asia). The second reason is that at higher latitudes advection, i.e., frontal systems, dominate weather formation, and as such were a priori believed to mask any influence of the land surface. However, a number of more recent studies suggest that also under mid- and high-latitude conditions the land surface can play a strong role in weather formation. Improvements in prescribing seasonal vegetation dynamics (Dirmeyer, 1994), as well as using better land surface (soil) schemes (Pielke *et al.*, 1997; Xue *et al.*, 1996, Betts *et al.*, 1996) improved weather prediction in the central US. Also, studies on anthropogenic land-cover changes in the same region suggested marked influences on (near surface) climate (Copeland *et al.*, 1996) and storm formation (Pielke, *et al.*, 1997). Results by Van der Hurk *et al.* (2001) indicate that also in Europe, at least in summer, land surface forcing of weather is significant. There are also indications of significant effects of land surface forcing at even higher latitudes. Betts and Ball (1997) showed how improvements in snow - albedo relationships, and in parameterising the influence of moss cover on soil evaporation enhanced the predictive capabilities of the ECMWF model.

Recently, also studies of global effects of historical land-use changes have emerged. Chase *et al.* (1996, 2000) and Zhao *et al.* (2000) found significant global temperature and precipitation changes, especially at higher latitudes, following a reasonably realistic pattern of vegetation change. Moreover, the relatively small perturbations imposed by Chase *et al.* (1996, 2000) appeared to tele-connect to cause large changes in temperature at higher latitudes and large changes in rainfall in the tropics. Zhao *et al.* (2000), using a different land-use change pattern and a different model configuration, support these results. For example, the seasonally averaged difference in temperature simulated by CCM3, resulting from a 17-year simulation with natural vegetation cover and a 17-year simulation using current land-cover, shows statistically significant differences in temperature, in some areas in the Northern Hemisphere reaching 3–4K.

Vegetation and soils play an important role as a store of carbon, while at the same time exchanging large quantities with the atmosphere (see also Chapter 5). Deforestation is the second major global source of atmospheric CO_2 after emission of carbon by fossil fuel combustion. The total sinks strength of the total terrestrial vegetation (Houghton *et al.*, 1996) is of comparable magnitude. This makes vegetation an important moderator of atmospheric CO_2 concentrations and thus an important aspect of greenhouse-gas related climate change. We discuss this in more detail in section 6 and only note here that atmospheric CO_2 concentrations also interact with plant physiology, further complicating the biospheric control of climate. The uptake of carbon dioxide by plants is strongly coupled to their regulation of evaporation. The regulation of stomatal conductance of plants is a function of (amongst others) their photosynthesis rate (see also Cox *et al.*, 1998). A number of studies indicate that increased carbon dioxide concentrations, resulting from fossil fuel combustion, may 'fertilise' plants in the sense that they grow more efficient while using less water (Walker *et al.*, 1997). This fertilisation could provide feedback on climate in two opposing ways.

This fertilisation may lead to a stronger sink strength, thereby mitigating atmospheric carbon dioxide increase and thus global warming. However, a larger uptake of CO_2 by the biosphere under higher CO_2 concentrations and a warmer climate would also mean at least a doubling of nitrogen fluxes into ecosystems (Schimel, 1998). As current nitrogen deposition trends do not indicate such increases, future C-cycling will probably be more limited by nitrogen.

This fertilisation could also imply that vegetation will sustain its growth rate at the expense of less water. Less evaporation implies smaller amounts of water are returned to the atmosphere, and at the same time a larger fraction of the sun's radiative energy is converted to warming the air. Accounting for this effect in climate change prediction studies results in increased global warming rates, comparable to ones due to the conventional greenhouse effect (Sellers *et al.*, 1996).

In the remainder of this chapter we will focus on land surface climate interactions in the two tropical regions mentioned before. First, we will review our current understanding of the possible role of land degradation in bringing about the severe droughts of the Sahel. Secondly, we will revisit the potential effects of tropical deforestation, in particular in the Amazon, on regional and global climate. In both regions, support by the NRP contributed to the Dutch participation in major field campaigns in these regions (HAPEX-Sahel, and in another desertification-threatened area EFEDA, and LBA respectively) and subsequent analysis of results. In this chapter, we will limit ourselves to a discussion of biosphere-atmosphere-hydrosphere interactions in terms of energy and water fluxes. Biogeochemical interactions are discussed more extensively in other sections of this book (Chapter 5).

2. Desertification in the Sahel

Since the 1960s the Sahel has experienced a drought that is unprecedented, at least for the past 1,600 years, in severity and persistence (Nicholson, 1979). The human dramas and socio-economic consequences resulting from drought-induced famines in the Sahel region provide a strong motivation for research into the causes of the drought. Since the early stages of this research, the causes for this prolonged drought have been sought somewhere between two extreme views. One view considers land surface degradation, resulting from population pressure in excess of the region's carrying capacity, as the main driver. The other view attributes the drought to unfavourable anomalous patterns in Sea Surface Temperature (SST) in the Atlantic ocean.

The Sahel is a bioclimatic zone of predominantly annual grasses with shrubs and trees, receiving a mean annual rainfall of between 150 and 600 mm per year. There is a steep gradient in climate, soils, vegetation, fauna, and land use, from the Sahara desert in the north to savannahs to the south. Climate and land-cover are very similar in the E-W direction, in contrast with a strong N-S gradient. The continental scale land mass and relative absence of orography (in the Western parts of the Sahel) allow for land surface-atmosphere interactions to play an important role in the regional climate. Furthermore, the Sahel summer climate is dominated by the West African monsoon system. The basic drive for the monsoon circulation is provided by the contrast in the thermal properties of the land and sea surfaces (Holton, 1992). Any changes in the contrast may substantially affect the monsoon flow. Finally, the region is located in the tropics, where dynamical flow instabilities are relatively weak (compared to mid-latitudes). The land surface determines the diabatic heating (latent and radiative heating), which in turn changes circulation and rainfall at seasonal and inter-annual time scales. Thus, land surface-atmosphere interactions play a major role in the regional climate and also make such interactions relatively easily detectable in model simulations.

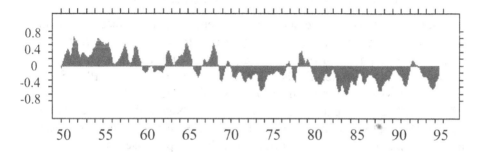

Figure 4.3. Rainfall index for the Sahel (8–20°N, 20W–10°E) 1950–1995. Positive values indicate above average rainfall, negative values indicate below average rainfall.

2.1 The drought of the last three decades

The rainy season in the Sahel develops with the northward movement of the Inter-tropical Convergence Zone (ITCZ), which causes humid air from the Gulf of Guinea to undercut the dry north-easterly air. West-moving squall lines and local convective activity cause mixing of the two air masses and results in short high-intensity rain events that decrease in frequency and rainfall amount towards the north where the humid air mass becomes shallower. Rainfall in the Sahel is characterised by high spatial and temporal variability, both within and between seasons, and by a high north-south gradient in the region of about 1 mm km^{-1} (Lebel *et al.*, 1992). Not only inter-annual variability is high, but also at decadal time scales dryer or wetter periods occur.

Since the late 1960s, a persistent summer drought lasted for more than 30 years. Although in the 1990s the rainfall was not as scarce as the 1980s, it was still below the climatological average (Figure 4.3). Rainfall appears to be redistributed along the north-south gradient. Rainfall in the Sahel has been reduced whereas rainfall to the south of 10°N increased. Barbe and Lebel (1997) have shown that the drought in Niger is caused by a decrease of the number of rain events at the height of the rainy season (July/August). The mean event rainfall and the length of the rainy season does not differ between drought (1970–1990) and pre-drought (1950–1970) years.

The continuous drought has also led to a substantial reduction in discharge of major rivers in the region (Savenije, 1995; Oki and Xue, 1998). For example, the discharge of the Niger River at Niamey was reduced by 34%, from an annual average of 1060 m^3 s^{-1} over the period 1929–1968 to 700 m^3 s^{-1} over the period 1969–1994.

Temperatures have been found to increase in the Sahel during dry years (Tanaka *et al.*, 1975; Schupelius, 1976). Observed temperature data show that summer surface temperatures over the Sahel region increased by 1–2 K during the summers of the 1980s compared with the 1950s (Xue, 1997).

2.2 Land atmosphere interactions in the Sahel

Experimental Studies Regional measurements of land surface energy and water balances, and of boundary layer and free-atmosphere dynamics are crucial in the development of realistic models. Two large field campaigns were conducted during the later 1980s (SEBEX) and 1991–1992 (HAPEX-Sahel) and another two are currently operational (SALT and CATCH).

The HAPEX-Sahel experiment took place in Niger, West Africa during 1991–1992 (Goutorbe *et al.*, 1994). The objective of this experiment was to improve the parameterisation of surface hydrological processes in semi-arid areas at scales consistent with general circulation models. The vegetation within the experimental area was typical of the southern Sahelian zone: arable crops (millet), fallow savannah and sparse dryland forest. The observations included a period of intensive measurement during the transition period of the rainy to the dry season, complemented by a series of long term measurements. Three super-sites were instrumented with a variety of hydrological and meteorological equipment to provide detailed information of surface energy exchange at the local scale. In addition, a network of automated weather stations was established over the full 1° square. Boundary layer measurements and aircraft measurements were also carried out to provide information about boundary layer development. The main experimental results have been presented in a Special Issue of the Journal of Hydrology (Goutorbe *et al.*, 1997).

The Dutch contribution (partially supported by NRP) focused mainly on surface-atmosphere fluxes of energy, water and CO_2, (Kabat *et al.*, 1997a; Jacobs and Verhoef, 1997, Lloyd *et al.*, 1997; Gash *et al.*, 1997; Moncrieff *et al.*, 1997a, 1997b) and on soil physical processes (Gaze *et al.*, 1997; Cuenca *et al.*, 1997) in the West Central Super Site (Kabat *et al.*, 1997b). Major progress was made with respect to methodological and technical aspects of eddy-correlation measurements, enabling continuous measurements of fluxes for several months (Moncrieff *et al.*, 1997b). This allowed detailed analysis of both spatial (Gash *et al.*, 1997) and temporal variability (Kabat et al., 1997a; Verhoef *et al.*, 1996) of energy fluxes, and of carbon fluxes (Moncrieff *et al.*, 1997a). Kabat *et al.* (1997a) showed how structurally very different vegetation types (savanah and 'tiger bush') had almost identical seasonal evaporation totals, while at the same time their sensible heat fluxes were quite different. Such measurements have been proven instrumental to the further development of surface energy balance models for this particular region, both stand-alone (Soet *et al.*, 2000) and as part of atmospheric models (Xue *et al.*, 1997; Hutjes *et al.*, 1997), both to quantify parameters and to further develop (sub-)models of specific processes.

GCM studies   Over the past twenty years, numerous studies with different models have been conducted examining the role of biospheric feedbacks in the Sahel drought of the same period. In the pioneering work by Charney *et al.* (1977) and confirmed by others (e.g., Chervin and Schneider, 1979; Sud and Fennessy, 1982) it was found that increases in albedo could lead to a reduction in precipitation. Later attention shifted more to the role of soil moisture and evapotranspiration (e.g., Walker and Rowntree, 1977; Sud and Fennessy, 1984). These modelling studies consistently demonstrated that the land surface may have a significant impact on the Sahel climate. These early studies employed simple surface layer models of the bucket type, only a single land surface parameter was tested each time, and the affected area and the magnitude of change were somewhat arbitrary.

A more realistic evaluation of the surface feedback to climate requires consideration of all comparable components of the energy and water balances. Therefore more recently, surface schemes which include more realistic representations of vegetation responses have been applied to the Sahel (e.g., Xue, 1997; Clark *et al.*, 2000). Some of these studies even attempted to reconstruct observed decadal climate anomalies.

Here we will review the study by Xue *et al.* (1997) in some more detail. It compares integrations with the Center for Ocean-Land-Atmosphere GCM and SSiB land surface model using a 'normal' vegetation map (control simulation) and several vegetation maps representing degraded land covers. Climatological sea surface temperatures (SST) were used as the lower atmospheric boundary conditions over the oceans. To average out the internal variability the model was integrated three times and ensemble means were used to detect climate impacts.

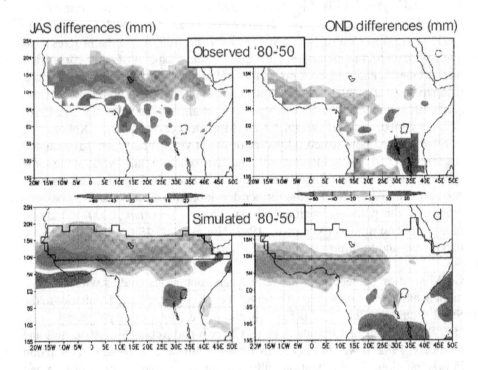

Figure 4.4. Observed (top) and simulated (bottom) Sahelian rainfall differences between the rainy season (left) of the 1980s and 1950s, as well as for the dry season (right) (adapted from Xue *et al.*, 1997).

Figure 4.4 shows the differences between the ensemble mean July-August-September (JAS) rainfall of the degraded and control simulations. Rainfall is reduced in the degraded area, but increases slightly to the south. This bi-modal

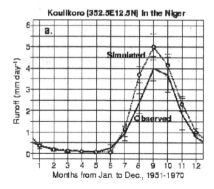
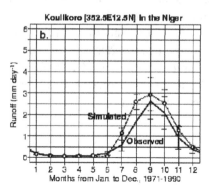

Figure 4.5. Observed and simulated river discharges in the Sahel for the drought period (right) and the pre-drought period (left) Oki and Xue (1998).

pattern is similar to observed rainfall patterns in dry periods. The rainfall changes are significant at the 90% confidence level in the areas enclosed by dotted lines. The simulated rainfall was found to be reduced by 39 mm per month, as compared to 45 mm per month observed. At the beginning of the Sahelian dry season (October-November-December, OND), when the ITCZ moves to the south, little rainfall is observed or simulated in the Sahel-proper. However, the areas of reduced rainfall related to land degradation move southward, while at the same time rainfall increases over eastern Africa, consistent with the observed OND rainfall. This suggests that the effect of land degradation is not limited to the rainy season within the Sahel, but extends to the autumn and to East Africa.

The same group analysed the impact of land-surface degradation on river discharge variability in tropical northern Africa (Oki and Sud, 1997; Oki and Xue, 1998). The simulated reduction in rainfall in degradation experiments led to reductions in simulated soil moisture, surface runoff, and subsurface drainage. To compare the runoff at grid points in GCM simulations with observed river discharge, a linear river routing scheme was applied with lateral flow directions taken from Total Runoff Integrating Pathways. Figure 4.5 compares simulated river discharge from control and degradation experiments at the Koulikoro station. This station represents the mean runoff from the Southwestern Sahel region. The mean monthly discharges are simulated fairly well and the contrast between control and degraded simulations corresponds to the observed difference between 1951–1970 and 1971–1990.

Clark *et al.* (2000) explored the relative importance of the location of vegetation degradation on its effects on rainfall. Five subregions were degraded in as many runs: northern Sahel, southern Sahel, West Africa, East Africa, and the coast area along the Gulf of Guinea. Considerable differences between the areas indicate that the location of degradation area is important. Degradation in the Northern Sahel and in the West African Sahel result in the largest and most significant reductions of rainfall. Degradation of the former area causes widespread reduction of rainfall across tropical northern Africa, both within and outside the degraded area. Deforestation in the coast area has marginal effects on rainfall.

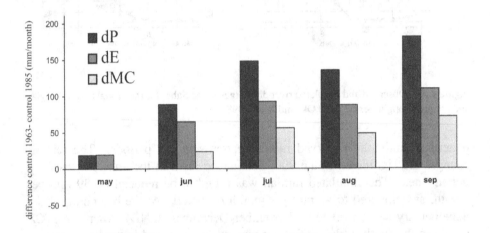

Figure 4.6.   Areal (5° W to 7° E, 13–17° N) differences in precipitation (dP - blue), evaporation (dE - purple) and moisture convergence (dMC - yellow) between 1963 and 1985, for the five rainy season months simulated. Units: mm per month.

Mesoscale studies   Mesoscale models used in Sahelian land surface- atmosphere interaction studies increase the level of physical detail, and their spatio-temporal resolution. Contrasts in surface energy partitioning have been found to be attributable to contrasts in soil moisture (Taylor *et al.*, 1997) and/or to contrasts in vegetation density (Hutjes *et al.*, 1999), and these models suggest that mesoscale heat flux gradients resulting from these contrasts are strong enough to influence rainfall patterns.

Hutjes and Dolman (2002) used the RAMS model to analyse the relative importance of SST anomalies and land surface changes on the rainfall production in the Sahel and to analyse possible mechanisms explaining these differences. After validating the model against HAPEX-Sahel observations for 1992, two runs with contrasting SST dynamics were compared (1963 – favourable to rainfall in the Sahel; 1985 – unfavourable), and two runs with contrasting vegetation cover (control as existed at the height of the drought, green as might have occurred prior to 1970). In 1963 rainfall is considerably (~40%) higher over the area than in 1985, with the main difference north of 12°N and east of 3°W where increases are very large.

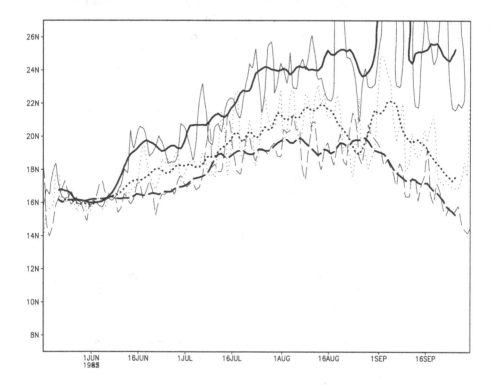

Figure 4.7. Seasonal progression of the ITCZ for 1992 (dash), 1985 (dotted) and 1963 (line). Shown is the daily noon-time position of the 7 g kg^{-1} contour of moisture at approx. 300m elevation (thin) and its ten day running mean (bold). This contour coincides with the 0 contour of north-south winds throughout the season.

Contrasting 1985 with 1963, Figure 4.6 shows that in the middle Sahel (13–17°N) the rainfall difference between 1963 and 1985 is particularly large in July, August and September. Evaporation and moisture convergence contribute in about a 60/40% ratio. These rainfall differences were found to be associated with different circulation patterns, consistent with Grist and Nicholson (2001). The Tropical easterly Jet (TEJ) in 1963 is stronger than in 1985, the African Easterly Jet (AEJ) weaker. But the main difference is found for the South Westerly Monsoon (SWM). Its winds are much stronger in 1963 and occur over a much deeper part of the atmosphere. At the same time also the humidity of this layer is much higher. These effects result in an ITCZ that moves about 3° to 4° further North in 1963 as compared to 1985 and 1992 respectively (Figure 4.7).

For both years 1963 and 1985 the control run, i.e., 'degraded' vegetation, was compared to the 'green' run, as shown in Figure 4.8. They show a difference between green and control for 1963 up to about 120 mm (or 20%). The area of maximum increase in rainfall appears to be displaced northward relative to the area of maximum increase in vegetation. In 1985 the relative difference is similar, though in absolute terms this amounts to about 40 mm only. In this case however the maximum increase in rainfall is more or less co-located with the area of maximum increase in vegetation.

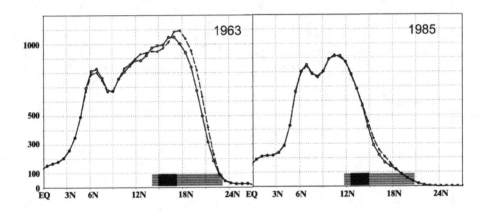

Figure 4.8. Zonal precipitation (averaged over lon 5W–7E) of green (dash) and control (line) runs for 1963 and 1985. The shaded box at the bottom indicates the zone with the most marked vegetation changes.

In general, the increase in evaporation in the 'green' run relative to the control run only partly leads to higher rainfall, as moisture divergence decreases following greening. In the mid-Sahel (13–17°N) evaporation in both years increases especially in the first three months, peaking in June, whereas later in the season the difference diminishes or even becomes negative. Comparing atmospheric circulation patterns between the 'green' and control runs reveals little difference except close to the surface, i.e., in the PBL.

This study suggests that land surface changes lead to differences in energy partitioning, with degradation decreasing PBL humidity but increasing low level convergence, resulting in a net relatively small increase in rainfall. SST differences on the contrary lead to modification of large scale circulation patterns stimulating moisture convergence (not only at low levels), and evaporation further enhancing precipitation.

Earth-system models of intermediate complexity　Over the last decade the prescribed vegetation characteristics in atmospheric models have been replaced in some models by dynamic vegetation models. This emerging field already produced a number of interesting results regarding land surface climate interactions in the Sahel. Some studies investigated the existence and stability of climate equilibrium states (e.g., Claussen and Gayler, 1997; Wang and Eltahir, 1999). Others focused more on simulation and understanding climate-vegetation variability on inter-annual to decadal time scales (e.g. Zeng *et al.*, 1999, 2000)

Using a GCM with an interactive vegetation model, Claussen and Gayler (1997) found multiple equilibrium states in climate/vegetation dynamics in northern Africa, which could not be found in other parts of the world. This discovery was confirmed by Wang and Eltahir (1999) with a coupled 2-dimensional climate/dynamic vegetation model. They found that the equilibrium vegetation/climate pattern was sensitive to the initial vegetation distribution. Starting with desert covering all of West Africa, the vegetation at equilibrium varied from tall grass near the coast to short grass and desert northward. In contrast, starting with forest all over west Africa, the equilibrium vegetation consisted mostly of forests covering most of West Africa with a narrow grassland band in the north, and a much higher productivity and rainfall than the first case. They also studied the effect of disturbances in model simulation. With less than 60% of the grass biomass removed between 12.5° and 17.5°N, the original vegetation and rainfall distribution could recover. With 60–75% biomass removed, the system converged to a new, 40% drier and 65% less productive equilibrium. When even more biomass was removed the system collapsed to a 60% drier and nearly 100% less productive equilibrium. These new drier states were associated with weaker large-scale circulation and suppressed local convection. They found however, that the process chain leading to this weakening/suppression differs between the wet and dry season. In the dry season, degradation led to higher albedo's, reduced net radiation, and cooler land surfaces. In the wet season degradation led to reduced evaporation and warmer land surfaces.

Zeng *et al.* (1999) could only reproduce the precipitation variability at decadal time scales in the Sahel if interactive vegetation processes were included in their model. In a companion paper Zeng *et al.* (2000) show how inter-annual variability in SST forcing affects the multiplicity of stable climate-vegetation patterns. Using the same SST climatology each year, again the equilibrium vegetation was sensitive to the initial vegetation distribution, consistent with both Claussen's and Wang's findings, reported above. However, using real SST's – exhibiting interannual variability on various time scales – to force the model, they found that equilibrium climate-vegetation patterns become almost independent of initial conditions.

Did land degradation cause desertification in the Sahel? Sahelian land-surface/atmosphere interactions have been studied for several decades now, starting with very simple models (Charney, 1975) that did not even include

hydrological processes. Later studies employed simple bucket type of land surface models coupled to GCMs, in turn expanded by vegetation models starting
from the late 1980s. The big experimental studies of the late 1980s and early
1990s added a lot in terms of understanding and quantitative information. All
this research showed one consistent result: the Sahelian climate is sensitive to
the local land-surface processes.

However, such a statement does not imply we can also confirm that land
degradation resulting from high population pressures is the cause for the drought
of the last decades. In principle, land-surface changes can result in rainfall reductions of the observed magnitude. If the land surface changes in the specified
control and desertification runs in Xue (1997) reflected real changes in vegetation properties between the 1980s and the 1950s, the results indicate that
degradation could lead to regional climate changes similar to those observed
between the 1980s and the 1950s. This concerns in particular the predicted increases in surface air temperature, and reductions in the summer rainfall, runoff,
and soil moisture over the Sahel region. Earlier studies attributed the drought
primarily to the albedo - radiation driven process chain. Now it is realised that
the evaporation - convection driven process chain is probably more important.
On the other hand the work by Hutjes *et al.* (2002) suggest vegetation changes
may only alter precipitation within constraints set by the external driver, i.e.,
sea surface temperature anomalies.

The use of dynamic vegetation models revealed another level of complexity. The CLIMBER studies (Claussen *et al.*, 1999) show how land surface
climate interactions may transform gradual external forcings (slow changes in
the earth's orbit around the sun) into much more stepwise changes in climate-
vegetation equilibria. And Zeng *et al.* (1999) show how SST anomalies may
be a primary driver of rainfall variability in the Sahel but that the increasingly
longer memories of soil moisture and vegetation are needed to reproduce realistic interannual precipitation variability.

At this stage, we still cannot decide whether the current drought is a temporary phenomenon, what the exact role of humans is in its cause, or what kind of
possible remedy should be taken. To assess the permanency of such droughts, at
least on human time scales, we may look back into the distant past (Claussen *et
al.*, 1999), or at other regions undergoing similar desertification trends such as
the Mediterranean, or Inner Mongolia (Xue, 1996). At the same time the type
of biophysical process studies reported here should be integrated with other
more human dimensions of the Sahelian drought, e.g., water resources, food
security issues, and socio-economic and political drivers of land-use change.
New process understanding may be generated, leading to better reconstruction
of past, and better predictions of future climate changes and anomalies in this
sensitive region. This should lead to better guidance in designing strategies to

adapt to, or perhaps even to mitigate drought and its consequences in the Sahel (see Chapter 7).

3.   Deforestation of the Amazon

The Amazon Basin contains the largest extent of tropical forest on earth, over 5×106 km^2. The annual discharge of the Amazon River into the Atlantic Ocean of more than 200,000 m^3 s^{-1} contributes about 18% of the global flow of fresh water into the oceans. The Amazon accounts for a large proportion of the planet's animal and plant species. In recent decades the Amazon is being deforested at a fast rate and with potentially large consequences for local and regional ecosystems, as well as for global water and carbon cycles and for atmospheric composition and functioning.

Humans have inhabited tropical South America probably for at least 12,000 years, but their impact on its ecosystems was quite small till a few decades ago. The impact of the several million indigenous people on the vegetation cover was negligible because they practiced small scale shifting and burning agriculturea-griculture. Species composition and distribution may have gradually changed, because many tribes selected useful plant species to grow near their dwellings, but this had little effect on vegetation cover. Even during the rubber boom most forests of Amazonia were left untouched and by the 1950s only about 1% of the forest of Amazonia had been cleared.

Since the 1950s the occupation and development of Amazonia has changed dramatically and is associated with a huge increase in extent and rate of deforestation. Government plans to develop and integrate the economy of Amazonia called for a network of roads criss-crossing the region and emphasised agricultural development.

In Brazil government incentives and tax benefits attracted large companies for cattle ranching, as well as millions of landless peasants from other parts of Brazil. As a result the population of Brazilian Amazonia has grown rapidly from 3,5 million in 1970 to almost 20 million in 2000, leading to the deforestation of over 500,000 km^2 in Brazil alone. Current rates of annual deforestation range from 15,000 to 20,000 km^2 in Brazilian Amazonia (INPE, 2000), see Figure 4.9. There are basically two spatial patterns of deforestation. Large cattle ranches produce rectangular-shaped (semi-) contiguous clearings ranging from a few km to 10 or 20 km on a side. Settling peasants produce so-called 'fish bone' patterns, that is, a patchy deforestation pattern along secondary roads 4 or 5 km apart.

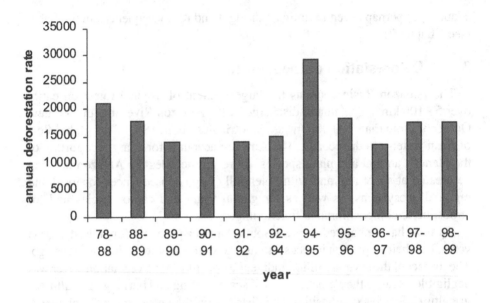

Figure 4.9.   Annual gross deforestation rates in Brazilian Amazonia (source INPE, 2000).

Observed land - atmosphere interactions in the Amazon   A number of field studies carried out over the last 15 years showed local changes in the water, energy, carbon and nutrient cycling, and atmospheric composition caused by deforestation and biomass burning. Large-scale and regional changes in the water cycle have not been detected so far. However, several GCM studies have been conducted on the potential impact of deforestation in the wet tropics. One of the first studies for the Amazon (Henderson-Sellers and Hornitz, 1984) predicted a 15–20 mm per month decrease in rainfall following a complete removal of tropical rainforest and its replacement by pasture. Later studies used improved land surface descriptions, based on measurements (Gash *et al.*, 1996). With such an improved land surface model coupled to the Hadley GCM, Lean and Rowntree (1997) predicted roughly 30 mm month^{-1} decrease in rainfall following complete conversion of forest to pasture. However, mesoscale studies with realistic, partial forest conversion suggested that with limited, relatively dispersed deforestation, rainfall reduction is much less severe and the rainfall might even increase, at least temporarily (IGBP, 1998). This is probably due to enhanced triggering of convection by contrasts in surface energy partitioning (forest vs. pasture), that occur in this particular area at a length scale similar to that at which convection is generated (compare Avissar and Liu, 1996, or Avissar, 1998).

Until recently, the interactions between Amazonian tropical forest and the atmosphere were poorly understood, especially in quantitative terms, because observations were largely missing due to the inaccessibility of the region. However the need to improve and validate the physics of the models that warned us of potentially catastrophic effects of deforestation has been the driving force behind a succession of land-atmosphere interaction studies in the region during the last 20 years.

Experimental Studies   The first observations of land surface - atmosphere exchanges measurements in Amazonia were made during the 1980s in the Amazon Regional Micrometeorology Experiment (ARME) (Shuttleworth, 1988). A collaboration between Brazilian and British scientists, ARME gathered data of the near surface climate, the proportion of rainfall intercepted and subsequently re-evaporated from the forest canopy, and the soil moisture status, continuously over a 25 month period at Reserva Ducke (Manaus). During four campaigns additional measurements of the surface energy balance, including radiation, and sensible and latent heat flux, vertical gradients of temperature, humidity and wind-speed, and plant physiological measurements were made (e.g., Shuttleworth *et al.*, 1984, 1988; Moore and Fisch, 1986; Roberts *et al.*, 1993). The ARME data were used extensively to calibrate models of e.g. surface conductance (Dolman *et al.*, 1991), GCM land surface sub-models (e.g., Sellers *et al.*, 1989), etc.

ARME concentrated on surface-atmosphere exchange processes. The Amazon Boundary Layer Experiment-2 (ABLE-2) focused on atmospheric transport and composition (Garstang *et al.*, 1990; and a special issue of the Journal of Geophysical Research, February 1988). A collaboration between US and Brazilian scientists ABLE 2 consisted of two phases: a dry season campaign (ABLE-2A, July–August 1985) and a wet season one (ABLE-2B, April to May 1987). During the wet season the air above the Amazon forest was found to be extremely clean, but during the dry season it was found to be heavily polluted as the result of biomass burning.

As a result of these experiments tropical forest had become relatively well parameterised, but the properties of ecosystems replacing forest were still poorly known. The Anglo-Brazilian Amazonian Climate Observation Study (ABRACOS) was set up to address this issue and also to extend the spatial and temporal scale of forest measurements. ABRACOS made measurements at paired forest-pasture sites in each of three locations, Manaus, Rondônia, in southwest Amazonia, and Marabá in the east (see Gash *et al.*, 1996). Continuous monitoring of climate and soil moisture was combined with campaigns involving micrometeorology (e.g. Wright *et al.*, 1995), plant physiology (e.g., McWilliam *et al.*, 1996) and detailed studies of the soil moisture changes (e.g., Tomasella and Hodnett, 1997). At the site in Rondônia three dry season atmospheric boundary layer sounding campaigns were carried out, known as the Rondônian Boundary Layer Experiment (RBLE; Nobre *et al.*, 1996). The results from RBLE showed that during the day the convective boundary layer grows to different depths over the two surface types as a result of contrasts in energy partitioning. It was also noted that cumulus clouds over pasture develop earlier in the day, than over the forest (Culf *et al.*, 1996).

Another extension of ABRACOS measured CO_2 fluxes at Rondônia (Grace *et al.*, 1995). These and subsequent observation showed the forest to be a strong sink for CO_2, which came as a big surprise at a time when the prevailing opinion expected forests to be in equilibrium (see also Chapter 5).

The currently implemented Large Scale Biosphere Experiment in Amazonia (LBA; Nobre *et al.*, 1996) differs from previous experiments by its thematic integration. It not only includes components focusing on meteorology and hydrology, but also on carbon, biogeochemistry, atmospheric chemistry and on the socio-economic drivers of environmental change in that particular region. Also it takes a long-term perspective with many observations being made continuously for multiple years. LBA is a Brazilian-led, international research initiative designed to create the new knowledge needed to understand the climatological, ecological, biogeochemical, and hydrological functioning of Amazonia, the impact of land-use change on these functions, and the interactions between Amazonia and the Earth system.

In the current context the Wet Atmospheric Mesoscale Campaign (WE-TAMC/LBA) (Silva Dias *et al.*, 2000), is probably the most successful component of LBA thus far. The WETAMC/LBA focused on both the local effects of deforestation and on the regional response to the larger scale forcing. The campaign was a joint venture between Brazilian and European scientists who joined forces with NASA's Tropical Rainfall Measurement Mission (TRMM) ground validation programme (Simpson *et al.*, 1996). The field phase took place in Rondônia during January and February 1999 and observations included: four high frequency radiosonde launch sites, four tethered balloons, two forest and four pasture flux measurement towers, a network of automatic weather stations, a dense rain gauge network, two doppler radars and two instrumented aircraft.

Dutch involvement in this type of Amazonian research, in part sponsored by the Dutch National Research Programme on Global Air Pollution and Climate Change, started with LBA. Currently still operational continuous monitoring of energy water and carbon fluxes at a number of sites (Andreae *et al.*, 2002; Araújo *et al.*, 2002; Kruijt *et al.*, 2001; von Randow *et al.*, 2002) is combined with intensive campaigns during e.g., the WET-AMC (Bink *et al.*, 2000abc; Silva Dias *et al.*, 2001, 2002).

Observed effects of deforestation   While already with ARME a number of surface characteristics were quantified for forest (e.g., albedo, rainfall interception dynamics, etc.), it was not until some years later that also cleared areas, covered with pasture, were monitored. Under the ABRACOS umbrella it was established that the pasture albedo ranged from 0.16 to 0.20 depending on the LAI, in contrast with forest albedos ranging from 0.11 to 0.13 correlating with soil moisture. Together with higher surface temperatures leading to more long-wave emissions this leads to a reduction of net radiation over pasture of about 11% compared to forests (Culf *et al.*, 1995).

Another marked contrast was found for evaporation, where forests show no seasonal trends, but pastures do. In the dry season the pastures exhibit moisture stress because its shallower roots cannot reach water, whereas forests may absorb water from depths greater than eight meters (Nepstad *et al.*, 1994). Forests do not appear to close stomata due to soil moisture stress, though they may do so in response to higher humidity deficits in the afternoon (Shuttleworth, 1988).

The WET-AMC has been very successful in illuminating the different rain-fall processes that occur in Amazonia, as well as the effects of deforestation and associated burning on these mechanisms. Rainfall in the Amazon displays more continental behaviour at the onset of the rainy season to more maritime characteristics at the height of the rainy season, though both regimes occur throughout the year (Silva Dias *et al.*, 2001). With large scale forcing, i.e., low level convergence associated with mid-latitude frontal systems and west-erly regimes, rainfall occurs throughout the day and clouds hardly grow above freezing levels and exhibit little lightning. In between such events, when no forcing is present and regimes are predominantly easterly, local convective systems develop which are more isolated in time and space, and which grow well above the $0°C$ isotherm and exhibit plenty of lightning (Rickenbach *et al.*, 2001). Such differences in cloud vertical structure and lightning intensity have been found in data from both ground based and TRMM satellite radars and lightning sensors. Also the diurnal cycle of rainfall shows differences between easterly and westerly regimes with the latter displaying a single peak around mid-afternoon, and the former with an additional minor peak in early morning (Marengo *et al.*, 2002). Finally, differences in cloud condensation nuclei (CCN) concentrations have been found between easterly ($\sim$800 cm^{-3}) and westerly regimes ($\sim$400 cm^{-3}) in the wet season, whereas in the dry season biomass burning leads to concentrations of up to $\sim$10000 cm^{-3}. Andrea *et al.* (2002) provide also evidence of the more oceanic character of much of the Amazon rainfall in terms of the chemical composition.

This leads us to a first effect of the very act of deforestation itself, burning, on rainfall production. Smoke causes a suppression of the warm rain processes based on collision and coalescence, as indicated by TRMM data showing that such clouds have to grow above the $-10°$ isotherm before they precipitate (Silva

Dias *et al.*, 2001c). In the wet season the CCN stem more from biogenic sources
of aerosols of more local origin (especially during easterlies).

Once pastures have been established other processes become relevant. Already during RBLE it was established that the PBL is warmer and grows more
rapidly to greater heights over pasture as compared to forest (Nobre *et al.*, 1996;
Fisch *et al.*, 1996). This is due to marked differences in energy partitioning between the two, with pastures having higher Bowen ratios (resulting from soil
moisture stress) and lower net radiation (higher albedo). In the wet season in
contrast the PBL on average does not differ between the two surfaces as also
the pastures are freely transpiring (Figure 4.10; Fisch *et al.*, 2002). The rapid
growth of the PBL over pasture in the dry season leads to earlier formation
of cumulus and a consequent reduction of solar irradiation (Culf *et al.*, 1996).
Cutrim *et al.* (1995) used satelite imagery to show that shallow cumulus formed
mostly over the cleared areas along the roads of Rondônia. In the wet season a
reverse situation may occur (Silva Dias *et al.*, 2000) where warmer night time
temperatures over forest lead to rainfall production from local moisture earlier
in the day than over pasture (Figure 4.11). In such cases the PBL over forest
may not grow as high as over the pasture due to the early onset of convective
downdrafts. To model such differences in timing is still a major challenge to
modellers as it depends not only on the parameterisation of processes such as
radiative transfer, PBL dynamics and surface energy partitioning but also on
their coupling (Silva Dias *et al.*, 2000).

3.1   Modelling land atmosphere interactions in the Amazon

Large scale effects of deforestation   Over the past decade deforestation experiments have been performed with most existing GCMs (Nobre *et al.*, 1991;
Dickinson and Kennedy, 1992; Lean and Rowntree, 1993, 1997; Polcher and
Laval, 1994ab; Manzi and Planton, 1996; Zang *et al.*, 1996; Hahmann and
Dickinson, 1997). Most of these assessed the possible impact of complete removal of Amazonian rainforest replacing it with pasture. In terms of surface
parameters this implied that albedo was changed from about 0.12 to 0.18, aerodynamic roughness from about 2 to 0.02 m, and evaporation parameters (LAI,
g_s) such that evaporation was reduced. In all but one the parameters were taken
from the ARME experiments, and only Lean and Rowntree (1997; see also
Figure 4.12) used ABRACOS data.

In all these deforestation studies net radiation was reduced as a consquence of
the imposed albedo increase, though partly compensated for by reduced cloud
cover, or by increased air temperatures leading to more downward longwave ra-

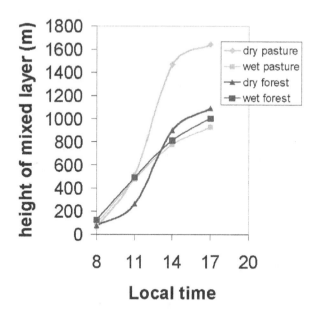

Figure 4.10. Observed differences in PBL height between forest and pasture in dry and wet season respectively (Fisch et al., 2002).

diation. In Manzi and Planton (1996), the albedo increase was even completely compensated for by this combination of feedbacks. Similarly, also all studies simulated a reduction in evaporation though the range of values was wide.

Some models simulated a nearly constant reduction throughout the year (e.g., Zang *et al.*, 1996), others found a maximum depression during the dry season (e.g., Lean and Rowntree, 1997), reflecting the ability of the land-surface scheme to represent properly the differences in soil moisture control on transpiration between pasture and forest and the differences in interception between the two. Surface temperatures were higher in all the deforested runs except two, where no significant changes of the annual mean were simulated.

Polcher and Avissar (2002) retraced these differences in model behaviour at the land-atmosphere interface to

(i) either differences in LSPs and the exact value of their parameters, or to

(ii) differences in atmospheric parameterisations (especially clouds) that feedback on the surface, or to the fact that

(iii) the control climate (e.g. seasonality) differed between these GCM's leading to different accents in their responses.

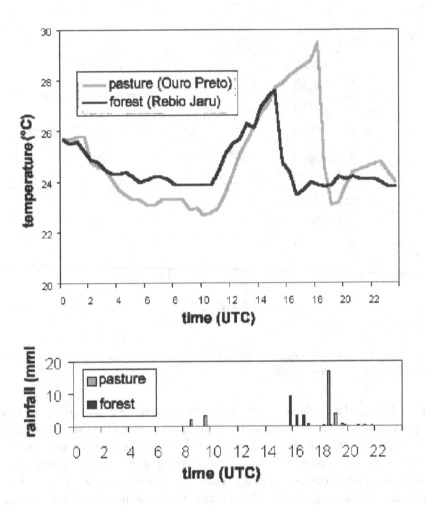

Figure 4.11.   Surface temperature and rainfall for pasture (Ouro Preto d'Oeste) and forest (Rebio Jaru) 7 february 1999.

The most striking disagreements between these GCM deforestation experiments occur with respect to impacts on moisture convergence. Polcher and
Laval, (1994b) found an increase in moisture convergence in contrast with all
previous studies. Later also Lean and Rowntree (1997) and Manzi and Planton
(1996) found increased moisture convergence, but still in contrast with contem-
porary work by Zang *et al.* (1996) and Hahmann and Dickinson (1997).

Tropical continental rainfall is characterised by short time scales and local
forcing, making it difficult to analyse their interactions with space-time averaged
data. Following Polcher and Avissar (2002) the evolution of the tropical con-

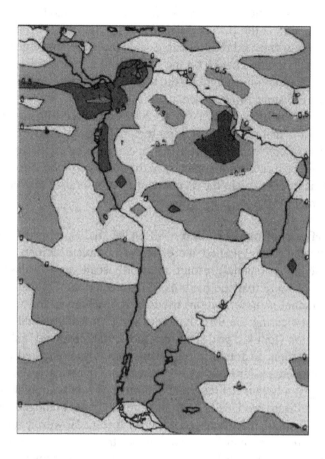

Figure 4.12. Annual rainfall (mm day^{-1}) differences between deforested and control runs with the Hadley Centre GCM (Lean and Rowntree, 1997).

vergence zone can be viewed as a rapid alternation between convective events, in which energy from the surface is carried upward and transformed, and situations of subsidence where the surface input is dominated by divergence in the lower layers. Polcher (1995) devised a method to separate such episodes based on the use of Potential Energy Divergence (PED) calculated for each gridbox on a daily basis. Thus analysing the GCM output from Polcher and Laval (1994b) they found that their precipitation change was associated with a reduction of the number of intense convective events (statistically significant during the months of May and June). The same study revealed that in situations of subsidence and weak convection the recycling rate of water was reduced, as direct consequence of the reduction of evaporation caused by deforestation. Polcher (1995) also showed that the number of convective events was very sensitive to the precise values of pasture parameters (varied within known uncertainties). The change in sensible heat flux at the surface determined the change in the number of intense convective events and thus the large scale moisture convergence over the region. The evaporation change on the other hand affected the recycling rate and thus the precipitation brought by the weaker convective events but without modifying the moisture convergence.

As pointed out by Polcher and Avissar (2002), an interesting consequence of this mechanism is that the total result on regional precipitation will depend on the distribution of convective and subsidence events. The climate of the control experiment will be a determining factor for the outcome of the deforestation experiment. This might explain the diversity of results for the other deforestation experiments mentioned.

Mesoscale effects of deforestation   All GCM studies reported above were unrealistic because they studied the effect of complete deforestation of the Amazon. The higher (spatial) resolution of mesoscale models allows analysis of the effects of more realistic patterns of deforestation. Dolman (1999, Wageningen, personal communication) used the RAMS model to study the effect of partial forest clearings on PBL dynamics and eventually rainfall production. In high resolution (four km grid size) runs the actual, observed 'fish-bone' pattern of deforestation, as found in e.g., Rondônia, was compared to complete forest and complete pasture land-cover scenarios. Given that a statistical proof could not be given because of the limited number of events simulated, some interesting results were produced. Rainfall simulated over homogeneous pasture was lower than simulated over homogeneous forest. However, rainfall simulated over the 'real landscape' was even higher than found over homogeneous forest, in line with the results by Cutrim *et al.* (1995) who showed that shallow cumulus preferentially developed over the deforested corridors in Rondônia.

This work built on previous work (Dolman *et al.*, 1999) which identified shortcomings of current mesoscale models to completely reproduce the observed differences in PBL growth over forest and pasture respectively. In line with Fisch *et al.* (1996) they were unable to simulate the observed rapid PBL heating and growth over the deforested area, while doing a good job in simulating PBL dynamics over forest. This was attributed to any or a combination of the following limitations: to a missing heat input from solar adsorption by aerosols, to the neglect of entrainment at the top of the PBL, and/or to a poor representation of the juxtaposition of small patches of forest and pasture leading to enhanced turbulence in the early morning causing a faster breakdown of the nocturnal inversion. Also using the RAMS model Silva Dias and Regnier (1996) showed that local circulations developed between pasture and surrounding forest in Rondônia during daytime in the dry season, which enhanced low level convergence leading to shallow cumulus development in spite of a deep mixed layer. Souza *et al.* (2000) show that not only the contrast in energy partitioning between forest and pasture induces different PBL temperatures and depths leading to local circulations, but that small elevation differences further enhance these circulations.

Deforestation and climate change   Combining the results as reported in the previous sections with projections of climate change due to global warming Nobre *et al.* (2002) made some interesting inferences about possible future scenarios for the Amazon. All IPCC projections of future climate, whether using low emission scenarios (550 ppm CO_2 by 2100) or high scenarios (830 ppm by 2100) indicate significant climate change for Amazonia. For Amazonia, projected temperature increases range from about 1°C (low emission scenario) to about 6°C (high emission scenario) by 2080 (Carter and Hulme, 2000). Precipitation changes by 0 to ±3% for the low emission scenarios and up to ±10% for the high emission scenarios, but some GCMs simulate decreases in precipitation where others simulate increases over most of Amazonia.

Such changes in precipitation are of the same magnitude as basin-wide rainfall changes associated with SST influences, or with large-scale deforestation. However, critical to the sustainability of tropical rain forests is the duration of the rainy season, which is yet harder to predict.

The uncertainty with respect the to sign of projected rainfall changes in the future makes it extremely difficult to 'predict' the impact of climate change on the Amazonia ecosystems (Nobre *et al.*, 2002). If there is reduction of rainfall, those reductions will add to those expected to take place as a response to large-scale deforestation. The result will be a significant increase in the susceptibility of the ecosystems to fire and reductions of species not tolerant to drought or fire (Nepstad *et al.*, 1994). On the other hand, if precipitation increases as a result of global warming, those increases might counteract the deforestation-induced rainfall reductions. The final result would be a smaller rainfall change. With respect to temperature changes, global warming would add to the temperature increase due to large-scale deforestation.

Such changes might trigger other feedbacks through its influence on carbon cyclinginxxcycle,carbon. Increased temperatures may increase heterotrophic respiration leading to less net carbon uptake by the forests or even net releases. Severe droughts might even cause forest die-backs that would release large amounts of carbon into the atmosphere over a period of just one to two decades. The CO_2 thus released would act as a positive feedback for the GHG radiative forcing (Cox *et al.*, 2000).

The uncertainty on how precipitation will change in Amazonia due to global warming prevents determining whether the final climatic feedback will be positive or negative, and renders the possibility of such natural disasters to acutely occur in the near or distant future as speculative at the moment. Hopefully, the results from projects like LBA will reduce the uncertainties in such predictions, and will suggest appropriate measures to mitigate such negative future scenarios.

4.   Concluding remarks

In this chapter we reviewed the coupling between the land surface and the climate as mediated by energy and water cycles. It was shown that such coupling can be strong. The surface characteristics, amount and type of vegetation, determine the partitioning of energy at the surface. This in turn affects heating and moistening rates of the planetary boundary layer, which affects cloud formation and the onset of convection, thereby also coupling to the free atmosphere. Such interactions cover a wide range of space and time scales making analysis of process chains far from trivial.

Nevertheless great progress has been made in recent decades, both in the observational and modelling domain. This has lead to considerable quantitative understanding of land surface-atmosphere interactions in different ecoclimatic regions of the world, and of the possible impacts of human alterations on the landscape on climate. Over the last few decades alterations in the landscape occur that have never before been as rapid or as extensive, in particular in the tropics. However, though many links between land-cover change and weather or climate change have been identified, conclusive statements about precise causes, or about the relative contribution of multiple causes cannot be made yet. It is even more difficult to predict in detail how such changes might be affected by other climate forcings, such as global warming and air pollution. This knowledge is urgently needed to devise policies that should make us less vulnerable to land-use change and induced climate change, either by adapting to existing or future situations, or by mitigating the more adverse causes and effects.

Acknowledgements

This paper reviews work of a large community. Structuring this synthesis for NRP has been a process overlapping with similar activities in other scientific networks. In particular, this compilation has benefited from work and discussions with Y. Xue, R.J. Harding, M. Claussen, S.D. Prince, E.F. Lambin, S.J. Alan, P.A. Dirmeyer, C. Nobre, M.A.F. Silva Dias, A.D. Culf, J. Polcher, J.H.C. Gash, J. Marengo, S. Denning, M.O. Andreae, P. Artaxo, R. Avissar.

References

Andreae, M.O., de Almeida, S.S., Artaxo, P., Brandão, C., Carswell, F. E., Ciccioli, P., Culf, A., Esteves, J.L., Gash, J.H.C., Grace, J. , Kabat, P., Lelieveld, J., Malhi, Y., Manzi, A.O., Meixner, F.X., Nobre, A.D., Nobre, C.A., de

Lourdes Ruivo, M.A., da Silva-Dias, M.A.F., Stefani, P., Valentini, R., von Jouanne, J. and Waterloo M.J. (2002) Towards an understanding of the biogeochemical cycling of carbon, water, energy, trace gases and aerosols in Amazonia: The LBA-EUSTACH experiments. Journal of Geophysical Research, Special issue LBA.

Araújo, A.C., Nobre, A.D., Kruijt, B., Culf, A.D., Stefani, P., Elbers, J.A., von Randow, C., Kabat, P., Mendes, D. (2002) The Manaus LBA site: new long-term studies of carbon dioxide fluxes over a Central Amazonian rain forest. Journal of Geophysical Research, Special issue LBA (in press).

Avissar, R. (1998) Which Type of SVATS is Needed for GCMs. Journal of Hydrology 212–213, 136–154.

Avissar, R. and Liu, Y. (1996) Three-dimensional numerical study of shallow convective clouds and precipitation induced by land surface forcing. Journal of Geophysical Research 101, 7499–7518.

Barbe, L. le, and Lebel, T. (1997) Rainfall climatology of the HAPEX-Sahel region during the years 1960–1990. Journal of Hydrology 188–189, 43–73.

Betts, A.K. and Ball, J.H. (1997) Albedo over Boreal Forest. Journal of Geophysical Research 102, 28901–28910.

Betts, A.K., Ball, J.H., Beljaars, A.C.M., Miller, M.J. and Viterbo, P. (1996) The land surface atmosphere interaction: a review based on observational and global modelling perpectives. Journal of Geophysical Research 101, 7209–7225.

Bink, N.J., Elbers, J.A., Holwerda, Kabat, P., Waterloo, M.J. (2000a) Energy balance measurements over pasture and Amazonian forest during the wet season. AMC LBA. LBA First Science Conference, Book of Abstracts, 228.

Bink, N.J., Elbers, J.A. Holwerda, Kabat, P., Waterloo, M.J. (2000b) Boundary layer observations in Rondônia with windprofiler, SODAR and RASS and tethersonde during the wet season. Atmospheric Measuring Campaign LBA. LBA First Science Conference, Book of Abstracts, 227.

Bink, N.J., Elbers, J.A., Kabat, P., Waterloo, M. (2000c) Carbon dioxide measurements over pasture and Amazonian forest during the wet season. AMC LBA. LBA First Science Conference, Book of Abstracts, 79.

Carter, T., and Hulme, M. (2000) Interim Characterizations of Regional Climate and Related Changes up to 2100 Associated with the Provisional SRES Marker Emissions Scenarios. IPCC Secretariat, c/o WMO, Geneva, Switzerland.

Charney, J.G. 1975. Dynamics of deserts and drought in the Sahel. Quarterly Journal of the Royal Meteorological Society 101, 193–202.

Chase, T.N., Pielke, R.A., Kittel, T.G.F., Nemani, R.R. and S.W. Running (1996) The sensitivity of a general circulation model to global changes in leaf area index. Journal of Geophysical Research, 101, 7393–7408.

Chase, T.N., Pielke, R.A., Kittel, T.G.F., Nemani, R.R. and S.W. Running (2000) Simulated impacts of historical land cover changes on global climate in northern winter. Climate Dynamics 16, 93–105.

Charney, J.G. (1975) Dynamics of deserts and drought in the Sahel. Quarterly Journal of the Royal Meteorological Society 101, 193–202.

Charney, J.G., Quirk, W.K., Chow S.-H. and Kornfield, J. (1977): A comparative study of the effects of albedo change on drought in semi-arid regions. Journal of Atmospheric Sciences 34, 1366–1385.

Chervin, R. M. and Schneider, S. H. (1979) On determining the statistical significance of climate experiments with general circulation models. Journal of Atmospheric Sciences 33, 405–412.

Clark, D., Xue, Y., Valdes, P. and Harding, R.J. (2000) Impact of land surface degradation over different subregions in Sahel on climate in tropical North Africa. Journal of Climate 14, 1802–1822.

Claussen, M. and Gayler, V. (1997) The greening of the Sahara during the mid-Holocene: results of an interactive atmosphere-biome model. Global Ecology and Biogeography Letters 6, 369–377.

Claussen, M., Kubatzki, C., Brovkin, V., Ganopolski, A., Hoelzmann, P. and Pachur, H.J. (1999) Simulation of an abrupt change in Saharan vegetation at the end of the mid-Holocene. Geophysical Research Letters 26, 2037–2040.

Copeland, J.H., Pielke, R.A. and Kittel, T.G.F. (1996) Potential climatic impacts of vegetations change: A regional modeling study. Journal of Geophysical Research 101, 7409–7418.

Cox, P.M., Huntingford, C. and Harding, R.J. (1998) A Canopy Conductance and Photosynthesis Model for use in GCM Land Surface Scheme, Journal of Hydrology 212–213, 79–94.

Cox, P.M., Betts, R.A., Jones, C.D., Spall, S.A. and Totterell, I.J. (2000) Acceleration of global warming due to carbon-cycle feedbacks in a coupled climate model. Nature 408, 184–187.

Cuenca, R.H., Brouwer, J., Cahnzy, A. , Droogers, P., Galle, S., Gaze, S.R., Sicot, M., Stricker, H., Angulo-Jaramillo, R., Boyle, S.A., Bromley, J., Chebhoumi, A.G., Cooper, J.D., Dixon, A.J., Fies, J.C., Gandah, M. , Gaudu, J.C., Laguerre, L., Lecocq, J., Soet, M., Stewart, H.J., Vandervaere, J.P., and Vauclio, M. (1997) Soil measurements during HAPEX-Sahel intensive observation period. Journal of Hydrology 188–189, 224–266.

Culf, A.D., Fisch, G. and Hodnett, M.G., 1995. The albedo of Amazonian forest and Rangeland. Journal of Climate 8, 1544–1554.

Culf, A.D., Esteves, J.L., Marques Filho, A. de O. and da Rocha, H.R. (1996) Radiation, temperature and humidity over forest and pasture in Amazonia. In: J.H.C. Gash, C.A. Nobre, J.M. Roberts, and R.L. Victoria (ed.), Amazonian Deforestation and Climate, John Wiley, Chichester, 175–191.

Cutrim, E., Martin, D.W. and Rabin, R. (1995) Enhancement of cumulus clouds over deforested lands in Amazonia. Bulletin of the American Meteorological Society 76, 1801–1805.

Dickinson, R.E. and Kennedy, P. (1992). Impact on regional climate of amazon deforestation. Geophysical Research Letters 19, 1947–1950.

Dirmeyer, P.A. (1994) Vegetation Stress as a feedback Mechanism in Mid-latitude Drought. Journal of Climate 7, 1463–1483.

Dolman, A.J., Gash, J.H.C., Roberts, J. and Shuttleworth, W.J. (1991) Stomatal and surface conductance of tropical rainforest. Agriculture and Forest Meteorology 54, 303–318.

Dolman, A.J., Dias, M.A.S., Calvet, J.C., Ashby, M., Tahara, A.S., Delire, C., Kabat, P., Fisch, G.A. and Nobre, C.A. (1999) Meso-scale effects of tropical deforestation in Amazonia: Preparatory LBA modelling studies. Annales Geophysicae, 1095–1110.

Fisch, G., Culf, A.D. and Nobre, C.A. (1996) Modelling convective boundary layer growth in Rondônia. In: J.H.C. Gash, C.A. Nobre, J.M. Roberts, and R.L. Victoria (ed.), Amazonian Deforestation and Climate, John Wiley, Chichester, 425–435.

Fisch, G., Tota, J., Machado, L.A.T., Silva Dias, M.A.F. da, Nobre, C.A., Dolman, A.J., Halverson, J., Fuentes, J.D., Culf, A.D. (2002) Convective Boundary layer over pasture and forest sites in Amazonia. Journal of Geophysical Research (in press).

Friedlingstein, P., Bopp, L., Ciais, P., Dufresne, J.-L., Fairhead, L., LeTreut, H., Monfray, P., Orr, J. (2001) Positive feedback between future climate change and the carbon cycle. Geophysical Research Letters 28, 1543–1546.

Garstang, M., Ulanski, S., Greco, S., Scala, J., Swap, R., Fitzjarrald, D., Browell, E., Shipman, M., Connors, V., Harriss, R. and Talbot, R. (1990) The Amazon Boundary-Layer Experiment (ABLE 2B): A meteorological perspective. Bulletin of the American Meteorological Society 71, 19–31.

Gash, J.H.C., Nobre, C.A., Roberts, J.M. and Victoria, R.L. (ed.) (1996) Amazonian Deforestation and Climate, John Wiley, Chichester.

Gash, J.H.C., Kabat, P., Monteny, B., Amadou, M., Bessemoulin, P., Billing, H., Blyth, E.M., de Bruin, H.A.R., Elbers, J.A., Friborg, T., Harrison, G., Holwil, C.J., Lloyd, C.R., Lhomme, J.P., Moncrieff, J.B., Puech, D., Soegaard, H., Taupin, J.D., Tuzet, A. and Verhoef, A. (1997) The variability of evaporation during the HAPEX-Sahel intensive observation period. Journal of Hydrology 188–189, 385–399.

Gaze, S.R., Simmonds, L.P., Brouwer, J. and Bouma, J. (1997) Measurement of surface redistribution of rainfall and modelling its effect on water balance calculations of millet field on sandy soils in Niger. Journal of Hydrology 188–189, 267–284.

Goutorbe, J.P.*et al.* (1994) HAPEX-SAHEL: a large scale study of land atmosphere interactions in the semi-arid tropics. Annales Geophysicae 12, 53–64.

Goutorbe, J.P., Dolman, A.J., Gash, J.H.C., Kerr, Y.H., Lebel, T., Prince, S.D. and Stricker, J.N.M. (Ed.) (1997) HAPEX - Sahel, Journal of Hydrology 188–189, Elsevier, Amsterdam, 1079.

Grace, J., Lloyd, J., McIntyre, J, Miranda, A., Meir, P., Miranda, H., Nobre, C., Moncrieff, J., Massheder, J., Malhi, Y., Wright, I. and Gash, J.H.C. (1995) Carbon dioxide uptake by an undisturbed tropical rain forest in South-West Amazonia 1992-1993. Science 270, 778–780.

Grist, J.P. and Nicholson, S.E. (2001) A study of the dynamic factors influencing the rainfall variability in the West-African Sahel. Journal of Climate 14, 1337–1359.

Hahmann, A.N. and Dickinson, R.E. (1997). RCCM2-BATS model over tropical South America: application to tropical deforestation. Journal of Climate 10(8), 1944–1964.

Henderson-Sellers, A., and Hornitz, (1984) Possible climatic impacts of land cover transformations with particular emphasis on tropical deforestation. Climate Change 6, 231–258.

Holton, J.R. (1992) An introduction to dynamic Meteorology. Academic Press, 507.

Houghton, J.T., Meiro-Filho, L.G., Callander, B.A., Harris, N., Kattenberg, A. and Maskell, K. (Ed.) (1996) Climate Change 1995 - The Science of Climate Change, Cambridge University Press, Cambridge, U.K.

Hurk, B.J.J.M., Dolman, A.J., Holtslag, A.A.M., Hutjes, R.W.A., van de Kasteele, J. Ronda, R. and Ijpelaar, R.J.M. (2001) The land component in the climate system. In: Berdowski, J.J.M., Guicherit, R. and Heij, B. (Ed.), The Climate System. Swets & Zeitlinger, Lisse, NL.

Hutjes, R.W.A., Kabat, P. and Dolman, A.J. (1997) Aggregated Land surface parameters. In: Kabat, P., Prince, S. D. and Prihodko (Ed.) Hydrologic Atmospheric Pilot Experiment in the Sahel (HAPEX-Sahel), Methods, Measurements and selected results from the West Central Supersite, Report 130, DLO Winand Staring Center for Integrated land soil and water research, Wageningen, the Netherlands, 279–285.

Hutjes, R.W.A. and Dolman, A.J. (1999) The effects of subgrid variability in vegetation cover and soil moisture on regional scale wheater: a case study for the Sahel. In: Harding *et al.* (ed.), Modelling the effect of Land degradation on Climate, Centre for Ecology and Hydrology, Wallingford, UK.

Hutjes, R.W.A., ter Maat, H.W. and Dolman, A.J. (2002) On the relative importance of landsurface versus SST forcing on seasonal rainfall in the Sahel. Journal of Geophysical Research, (submitted).

IGBP (1998) LBA is moving forward. Global Change Newsletter 33, 1–9.

Jacobs, A.F.G. and Verhoef, A. (1997) Soil evaporation from sparse natural vegetation estimated from Sherwood numbers. Journal of Hydrology 188–189, 443–452.

INPE (2000) http://sputnik.dpi.inpe.br:1910/col/dpi.inpe.br/lise/2001/05.16.09.55/ doc/html/pag_8.htm.

Kabat, P., Dolman, A.J. and Elbers, J.A. (1997a) Evaporation, sensible heat and canopy conductance of fallow savannah and patterned woodland in the Sahel. Journal of Hydrology 188–189, 494–515.

Kabat, P., Prince, S.D. and Prihodko (Ed.) (1997b): Hydrologic Atmospheric Pilot Experiment in the Sahel (HAPEX-Sahel), Methods, Measurements and selected results from the West Central Supersite, Report 130, DLO Winand Staring Center for Integrated land soil and water research, Wageningen, the Netherlands.

Kruijt, B., Elbers, J.A., von Randow, C., Araújo, A.C., Culf, A., Bink, N.J., Oliveira, P.J., Manzi, A.O., Nobre, A.D., Kabat, P. (2001). Aspects of the robustness in eddy correlation fluxes for Amazon rainforest conditions. Ecological Applications, submitted.

Lean, J. and Rowntree, P.R. (1996) Understanding the sensitivity of a GCM simulation of Amazonian deforestation to the specification of vegetation and soil characteristics. Climate Research Technical Note 65, Hadley Centre, UK.

Lean, J. and Rowntree, P. (1997) Understanding the sensitivity of a GCM simulation of Amazonian deforestation to the specification of vegetation and soil characteristics. Journal of Climate, 10, 1216–1235.

Lean, J. and Rowntree, P. (1993). A GCM simulation of the impact of Amazonian deforestation on climate using an improved canopy representation. Quarterly Journal of the Royal Meteorological Society 119, 509–530.

Lebel, T., Sauvageot, H., Hoepffner, M., Desbois, M., Guillot, B. and Hubert, P. (1992) Rainfall estimation in the Sahel: the EPSAT-Niger experiment. Hydrological Sciences Journal 37, 201–215.

Lloyd, C.R., Bessemoulin, P., Cropley, F.D., Culf, A.D. , Dolman, A.J., Elbers, J., Heusinkveld, B., Moncrieff, J.B., Monteny, B. and Verhoef, A. (1997) A comparison of surface fluxes at the HAPEX-Sahel fallow bush sites. Journal of Hydrology 188–189, 400–425.

Manzi, A.O. and Planton, S. (1996) Calibration of a GCM using ABRACOS and ARME data and simulation of Amazonian deforestation. In: J.H.C. Gash, C.A. Nobre, J.M. Roberts, and R.L. Victoria (ed.), Amazonian Deforestation and Climate, John Wiley, Chichester, 505–529.

Marengo, J., Fisch, G., Vendrame, I., Cervantes, I., Morales, C. (2002) On the diurnal and day-to-day variability of rainfall in Southwest Amazonia during the LBA-TRMM and LBA-WET AMC campaigns of summer 1999. Journal of Geophysical Research, Specisal issue on LBA (in press).

McWilliam, A.L.C., Cabral, O.M.R., Gomes, B.M., Esteves, J.L., Roberts, J. (1996) Forest and pasture leaf-gas exchange in south-west Amazonia. In: J.H.C. Gash, C.A. Nobre, J.M. Roberts, and R.L. Victoria (ed.), Amazonian Deforestation and Climate, John Wiley, Chichester, 265–285.

Moncrieff, J.B., Monteny, B., Verhoef, A., Friborg, T., Elbers, J.A., Kabat, P., de Bruin, H.A.R., Soegaard, H., Jarvis, P.G. and Taupin, J.D. (1997a) Spatial and temporal variations in net carbon flux during HAPEX-Sahel. Journal of Hydrology 188–189, 563–558.

Moncrieff, J.B., Massheder, J.M., de Bruin, H.A.R., Elbers, J.A., Friborg, T., Heusinkveld, B., Kabat, P., Scott, S., Soegaard, H. and Verhoef, A. (1997b) A system to measure surface fluxes of momentum, sensible heat, water vapor and carbon dioxide. Journal of Hydrology 188–189, 559–611.

Moore, C.J. and Fisch, G. (1986) Estimating heat storage in Amazonian tropical forest. Agriculture and Forest Meteorology 38, 147–169.

Nepstad, D.C., Carvalho, C.R., Davidson, E.A., Jipp, P.H., Lefebvre, P.A., Negrelros, G.H., da Silva, E.D., Stone, T.A., Trumbore, S.E. and Vieira, S. (1994) The role of deep roots in the hydrological and carbon cycles of Amazon forests and pastures. Nature 372, 666–669.

Nicholson, S. E. (1979) Revised rainfall series for the west African subtropics. Monthly Weather Review 107, 620–623.

Nobre, C., Sellers, P., and Shukla, J. (1991). Amazonian deforestation and the regional climate change. Journal of Climate 4, 957–988.

Nobre, C.A., Fisch, G., da Rocha, H.R., Lyra, R.F., da Rocha, E.P. and Ubarana, V.N. (1996) Observations of the atmospheric boundary layer in Rondônia. In: J.H.C. Gash, C.A. Nobre, J.M. Roberts, and R.L. Victoria (ed.), Amazonian Deforestation and Climate, John Wiley, Chichester, 413–424.

Nobre, C.A., Dolman, A.J., Gash, J.H.C., Hutjes, R.W.A., Jacob, D.J., Janetos, A.C., Kabat, P., Keller, M., Marengo, R.J., McNeal, R.J., Mellilo, J., Sellers, P.J., Wickland, D.E. and Wofsy, S.C. (ed.) (1996) The Large Scale Biosphere Atmosphere Experiment in Amazonia (LBA): Concise experimental plan, DLO Staring Centre for Integrated Land Soil and Water Research, Wageningen, Netherlands.

Nobre, C., Silva Dias, M.A.F., Kabat, P., Culf, A.D., Polcher, J., Gash, J.H.C., Marengo, J., Denning, S., Andreae, M.O., Artaxo, P., Avissar, R. (2002) The Amazonian climate. In: Kabat, P., Claussen, M., Dirmeyer, P.A., Gash, J.H.C., Bravo de Guenni, L., Meybeck, M., Pielke, R.A. (Sr.), Vörösmarty, C.J., Hutjes. R.W.A. and Lutkemeier, S., Vegetation, Water, Humans and the Climate: a new perspective on an interactive system, Springer Verlag, (in press).

Oki, T., and Sud, Y.C. (1997) Design of Total Runoff Integrating Pathways (TRIP)-A Global River Channel Network. Earth Interactions 2.

Oki, T. and Xue, Y. (1998) Investigation of river discharge variability in Sahel desertification experiment. Preprint of Ninth Symposium on Global Change Studies, 259–260.

Pielke, R.A., Lee, T.J., Copeland, J.H., Eastman, J.L, Ziegler, C.L, and Finley, C.A. (1997) Use of USGS provided data to improve weather and climate simulation. Ecological Applications 7, 3–21.

Pielke, R.A., Avissar, R., Raupach, M., Dolman, A.J., Zeng, X. and Denning, S. (1998) Interactions between the atmosphere and terrestrial ecosystems: Influence on weather and climate. Global Change Biology 4, 461–475.

Polcher, J. (1995). Sensitivity of tropical convection to land surface processes. Journal of Atmospheric Sciences 52 (17), 3143–3161.

Polcher, J. and Avissar, R. (2002) Deforestation and Climate. In: Kabat, P., Claussen, M., Dirmeyer, P.A., Gash, J.H.C., Bravo de Guenni, L., Meybeck, M., Pielke, R.A. (Sr.), Vörösmarty, C.J., Hutjes. R.W.A. and Lutkemeier, S., Vegetation, Water, Humans and the Climate: a new perspective on an interactive system, Springer Verlag, (in press).

Polcher, J. and Laval, K. (1994a) The impact of African and Amazonian deforestation on tropical climate. Journal of Hydrology 155, 389–405.

Polcher, J. and Laval, K. (1994b) A statistical study of regional impact of deforestation on climate of the LMD-GCM . Climate Dynamics 10, 205–219.

von Randow, C.S., L.D.A., Prasad, G.S.S.D., Manzi, A., Kruijt, B. (2002) Scale variability of surface fluxes of energy and carbon over a tropical rain forest in South-West Amazonia, I Diurnal Conditions. Journal of Geophysical Research, Special issue LBA (in press).

Rickenbach, T.M., Nieto Ferreira, R., Silva Dias, M.A.F. and Halverson, J. (2001) Modulation of convection in western Amazon basin by extratropical baroclinical waves. Journal of Geophysical Research, Special issue LBA (submitted).

Roberts, J.M., Cabral, O.M.R., Fisch, G., Molion, L.C.B., Moore, C.J. and Shuttleworth, W.J. (1993) Transpiration from Amazonian rainforest calculated from stomatal conductance measurements. Agriculture and Forest Meteorology 65, 175–196.

Savenije, H.H.G. (1995) New definitions for moisture recycling and the relationship with land-use changes in the Sahel, Journal of Hydrology 167, 57–78.

Schimel, D.S. (1998) Climate change: the carbon equation. Nature 393, 208–209.

Schupelius, G.D. (1976) Monsoon rains over west Africa. Tellus 28, 533.

Sellers, Bounoua, Collatz, Randall, Dazlich, Los, Berry, Fung, Tucker, Field, Jensen (1996) Comparison of Radiative and Physiological Effects of Doubled Atmospheric CO_2 on Climate. Science 271, 1402–1406

Sellers, P.J., Shuttleworth, W.J., Dorman, J.L., Dalcher, A., Roberts, J. M. (1989) Calibrating the simple bio-sphere model for Amazonian tropical forest using field and remote sensing data: Part1, Average calibration with field and remote sensing data. Journal of Applied Meteorology 28, 727–759.

Shuttleworth, W.J., Gash, J.H.C., Lloyd, C.R., Roberts, J.M., Marques, A de O., Fisch, G., de Silva, P., Ribeiro, M.N.G., Molion, L.C.B., de Abreu Sa, L.D., Nobre C.A., Cabral, O.M.R., Patel, S.R. and de Moraes, J.C. (1984) Eddy correlation measurements of energy partition for Amazonian forest. Quarterly Journal of the Royal Meteorological Society 110, 1143–1162.

Shuttleworth, W. J. (1988) Evaporation from Amazonian rainforests. Philosophical Transactions of the Royal Society B233, 321–346.

Silva Dias, M.A. F. and Regnier, P. (1996) Simulation of Mesoscale Circulations in a Deforested Area of Rondônia in the Dry Season. In: J.H.C. Gash, C.A. Nobre, J.M. Roberts, and R.L. Victoria (ed.), Amazonian Deforestation and Climate, John Wiley, Chichester, 531–547.

Silva Dias, M. A. F., Rutledge, S., Kabat, P., Silva Dias, P.L., Nobre, C., Fisch, G., Dolman, A.J., Zipser, E., Garstang, M., Manzi, A., Fuentes, J.D., Rocha, H., Marengo, J., Plana-Fattori, A., S , L., Alval , R., Andreae, M.O., Artaxo, P., Gielow, R. and Gatti, L. (2002) Clouds and rain processes in a biosphere atmosphere interaction context in the Amazon Region. Journal of Geophysical Research, Special issue LBA (in press).

Silva Dias, M.A.F., Nobre, C. and Marengo, J. (2001). The Interaction of clouds and rain with the biosphere. Global Change Newsletter 45, 8–11.

Silva Dias, M.A.F., Dolman, A.J., Rutledge, S., Zipser, E., Silva Dias, P., Fisch, G., Nobre, C., Kabat, P., Ferrier, B., Betts, A., Halverson, J., Garstang, M., Fuentes, J., Manzi, A., Rocha, H., Marengo, J.A., Morales, C., Bink, N.J. (2002) Convective systems and surface processes in Amazonia during the WETAMC/LBA. Journal of Geophysical Research, Special issue LBA (in press).

Simpson, J., Kummerow, C., Tao, W.K. and Adler, R.F. (1996) On the tropical rainfall measuring mission (TRMM). Meteorology and Atmospheric Physics 60, 19–36.

Soet, M., Ronda, R.J., Stricker, J.N.M. and Dolman, A.J. (2000) Land surface scheme conceptualisation and parameter values for three sites with contrasting soils and climate. Hydrology and Earth System Sciences 4, 283–294

Souza, E.P., Rennó, N.O., Silva Dias, M.A.F. (2000) Convective circulations induced by surface heterogeneities. Journal of Atmospheric Sciences 57, 2915–2922.

Sud, Y. C. and Fennessy, M. (1982) A study of the influence of surface albedo on July circulation in semiarid regions using the GLAS GCM. Journal of Climatology 2, 105–125.

Sud, Y.C. and Fennessy, M. (1984) A numerical study of the influence of evaporation in semiarid regions on the July circulation. Journal of Climatology 4, 383–398.

Tanaka, M., Weare, B. C., Navato, A. R. and Newell, R.E. (1975) Recent African rainfall patterns. Nature 255, 201.

Taylor, C.M., Harding, R.J., Thorpe, A.J. and Bessemoulin, P. (1997). A mesoscale simulation of land surface heterogeneity from HAPEX-Sahel. Journal of Hydrology 188–189, 1040–1066.

Tomasella, J. and Hodnett, M.G. (1997) Estimating soil water retention characteristics from limited data in Brazilian Amazonia. Soil Science 163, 190–202.

Verhoef, A., Allen , S.J., de Bruin, H.A.R., Jacobs, C.J.M. and Heusinkveld, B. (1996) Fluxes of water vapour and carbon dioxide from a Sahelian savannah. Agricultural and Forest Meteorology 80, 231–248

Walker, B., Steffen, W., Bondeau, A., Bugmann, H., Campbell, B., Canadell, P., T., C., Cramer, W., Ehleringeer, J., Elliot, T., Foley, J., Gardner, B., Goudriaan, J., Gregory, P., Hall, D., T., H., Ingram, J., Krner, C., Landsberg, J., Langridge, J., Lauenroth, B., Leemans, R., Linder, S., McMurtrie, R., Menaut, J.C., Mooney, H., Murdiyarso, D., Noble, I., Parton, B., Pitelka, L., Ramakrishnan, K., Sala, O., Scholes, B., Schulze, D., Shugart, H., Stafford-Smith, M., Suthurst, B., Valintin, C., Woodward, I. and Zhang, X.S. (1997) The terrestrial bioshere and global change. Implications for natural and managed ecosystems. A synthesis of GCTE and related research. IGBP, Stockholm.

Walker, J. and Rowntree, P.R. (1977) The effect of soil moisture on circulation and rainfall in a tropical model. Quarterly Journal of the Royal Meteorological Society 103, 29–46.

Walker, J. and Rowntree, P.R. (1999) Biosphere-atmosphere interactions over west Africa. 2. Multiple Climate Equilibria. Quarterly Journal of the Royal Meteorological Society 126, 1261–1280.

Wang, G. and Eltahir E.A.B. (1999) Biosphere-atmosphere interactions over west Africa. I: Development and validation of a coupled dynamic model, Quarterly Journal of the Royal Meteorological Society 126, 1239–1260.

Williams, E. *et al.* (2001) Observational tests of the aerosol hypothesis for precipitation and electrification in Rondônia, Brazil. Journal of Geophysical Research submitted.

Wright, I.R., Manzi, A.O., da Rocha, H.R., (1995) Canopy surface conductance of Amazonian pasture: model application and calibration for canopy climate. Agriculture and Forest Meteorology 75, 51–70.

Xue, Y. (1996) The impact of desertification in the Mongolian and the Inner Mongolian Grassland on the regional climate. Journal of Climate 9, 2173–2189.

Xue, Y. (1997) Biosphere feedback on regional climate in tropical north Africa. Quarterly Journal of the Royal Meteorological Society 123 B, 1483–1515.

Xue, Y., Elbers, J.A., Zeng, F.J. and Dolman, A.J. (1997) GCM parameterisation for Sahelian land surface proceses. In: Kabat, P., Prince, S. D. and Prihodko (Ed.) Hydrologic Atmospheric Pilot Experiment in the Sahel (HAPEX-Sahel), Methods, Measurements and selected results from the West Central Supersite, Report 130, DLO Winand Staring Center for Integrated land soil and water research, Wageningen, The Netherlands, 289–296.

Xue, Y., Fennessy, M.J. and Sellers, P.J. (1996) Impact of vegetation properties on US weather prediction. Journal of Geophysical Research 101, 7419–7430.

Zang, H., Henderson-Sellers, A. and McGuffie, K. (1996) Impacts of tropical deforestation. part I: Process analysis of local climate change. Journal of Climate 9, 1497–1517.

Zeng, N., Neelin, J.D., Lau, W. K.-M. and Tucker, C.J. (1999) Vegetation-climate interaction and Sahel climate variability. Science 286, 1537–1540.

Zeng, N., Neelin, J.D. (2000) The Role of Vegetation-Climate Interaction and Interannual Variability in Shaping the African Savanna. Journal of Climate 13, 2665–2670.

Zhao, M., Pitman, A.J. and Chase, T. (2000) The impact of land cover change on the atmospheric circulation. Climate Dynamics 17, 467–477.

Chapter 5

LAND USE AND TERRESTRIAL CARBON SINKS

A.J. Dolman
Vrije Universiteit Amsterdam

G.J. Nabuurs
Alterra, Wageningen University and Research Centre

P.K. Kuikman
Alterra, Wageningen University and Research Centre

R.W.A. Hutjes
Alterra, Wageningen University and Research Centre

J. Huygen †
Alterra, Wageningen University and Research Centre

A. Verhagen
Plant Research International, Wageningen University and Research Centre

L.M. Vleeshouwers
Plant Research International, Wageningen University and Research Centre

1.   Introduction: the global carbon cycle

Measurements of atmospheric CO_2 during the 1980s suggest that from the estimated 7.1 Gt C released by man (5.4±0.3 Gt C from fossil fuel and 1.7±0.9 Gt C from land-use change and deforestation) only 3.3±0.1 Gt C is found back in the atmosphere. From this amount 1.9±0.6 Gt C is absorbed by the oceans. The remaining 1.7 Gt is what is commonly and misleadingly referred to as

A.J. Dolman et al. (eds.), Global Environmental Change and Land Use, 111-136.
© 2003 *Kluwer Academic Publishers.*

the missing sink, and is increasingly allocated to the terrestrial surfaces of the Northern hemisphere (e.g., Bousquet *et al.*, 2000; Fan *et al.*, 1998).

The debate as to where the sink is located and how variable it is in time, and, in particular what is causing it, is still continuing. Amidst these scientific controversies, 172 countries signed the UNFCCC Kyoto protocol, in which the use of sinks is allowed in an effort to curb the continuing increase of CO_2 in the atmosphere. There is considerable scientific uncertainty on how to include the sinks in the protocol. This is to a large extent caused by the relatively poor knowledge of the global carbon cycle and the effects of management on the amount of carbon sequestered by the land surface. This chapter aims to provide a broad overview of that knowledge. It starts doing so by describing the global carbon balance and its components. Subsequently methods are described that allow us to measure components of the carbon balance. We then describe fluxes in agricultural and non-forest land and the interannual variability of the exchange of carbon between the atmosphere and land surface. The chapter ends by providing a first look at a method of designing gridded carbon inventories of Europe and a discussion on the policy relevance of the work described in this chapter.

The global terrestrial carbon budget consists of two large, almost balancing numbers: uptake by photosynthesis and release by respiration and decomposition of dead matter. It is important to define the various terms in the terrestrial carbon balance, as often, different measurement techniques refer to different terms of the carbon balance (Figure 5.1). Plants take up CO_2 from the air and convert it into carbohydrates by using photosynthetically active radiation (PAR). This process is called photosynthesis and translates into Gross Primary Production (GPP). Globally this term is about 120 Gt C yr^{-1}. Plants also respire for their maintenance and the sum of photosynthesis and autotrophic plant respiration (R_a) is called Net Primary Production (NPP) and amounts globally to about 60 Gt C yr^{-1}. Mortality of leaves and twigs produces dead organic material that is decomposed (SOM). The resulting term incorporating this decomposition, heterotrophic respiration (R_h), is Net Ecosystem Production (NEP) and is considerably smaller than NPP, about 10 Gt C yr^{-1} globally. Net Biome Production (NBP) is the long-term carbon uptake, including disturbance cycles of harvest, forest fires (producing black carbon) and other disturbances such as pests. This is the long term carbon storage and amounts to 0.7 Gt C yr^{-1} globally. The difference between GPP and NBP is orders of magnitude, so some care in the interpretation of the various results has to be taken.

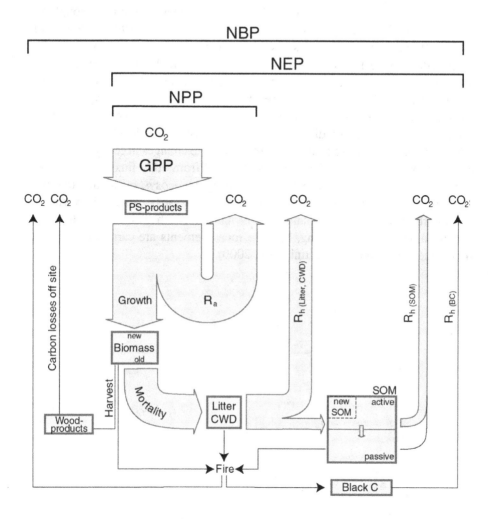

Figure 5.1. The various components of the terrestrial carbon balance. See text for explanation, redrawn after Schulze *et al.* (2000).

2. How to estimate the CO_2 uptake of the land?

Over the past three decades several methods have been developed and applied to quantify terrestrial carbon sources and sinks. Each of these methods has its individual strengths and weaknesses. These methods include inversions based on atmospheric chemistry (Bousquet *et al.*, 2000), biogeochemical models of the ecosystem (Schimel *et al.*, 2000), land-use bookkeeping models (Houghton *et al.*, 1999), CO_2-flux measurement towers (Martin *et al.*, 1998; Valentini *et al.*, 2000) and forest inventories. While atmospheric inversions constrain the magnitude of terrestrial carbon sinks, they have limited ability to discern the responsible mechanisms or exact location of the observed sink. Global bio-geochemical models can explore the importance of ecosystem physiological responses to climate variability or increasing CO_2, but they do not yet consider natural or human-induced disturbances. In contrast, methods that focus on the effects of human land-use changes are insensitive to changes in ecosystem phys-iology (Houghton *et al.*, 1999). Measurements from eddy flux towers reflect one signal from all of the mechanisms affecting net ecosystem production, but these local measurements at a few sites do not necessarily capture the variabil-ity of carbon flux across the landscape or nation. Neither do they capture the human influence as harvesting, because measurements are carried out over a short time period only (Valentini *et al.*, 2000).

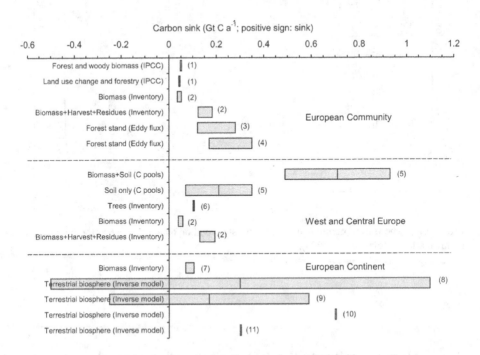

Figure 5.2. Estimates of the carbon sink in the European biosphere (Gt C yr^{-1}). Note that these refer to several techniques and to different geographical and political entities. (1) EEA/ETC Air Emissions (1999); (2) Kauppi and Tomppo (1993); (3) Martin (1998); (4) Martin *et al.* (1998); (5) Schulze *et al.* (2000); (6) Nabuurs *et al.* (1997); (7) Kauppi *et al.* (1992); (8) Bousquet *et al.* (1999); (9) Kaminski *et al.* (1999); (10) Rayner *et al.* (1997); (11) Ciais *et al.* (1995).

All of these different methods have produced a variety of estimates on the location and timing of the terrestrial carbon sink. Figure 5.2 displays this variety – grouped by method – for the sink estimates for the European land area. This figure is an accurate description of our ability to determine the sinks strength of large areas, the sizes of a continent. There are large differences in the estimates obtained by various methods. Note also that some of these refer to political Europe and others to geographical Europe. All methods estimate a positive sink, none shows a net source or no uptake. However a factor 5 appears between the estimates obtained by Schulze (2000) and those of forest inventories (Nabuurs *et al.*, 1997). All the estimates are close to the estimate of Houghton *et al.* (1999) found for North America for land use change for forestry. A key problem however, is that the various methods use different scaling techniques to transform information obtained at one spatial and temporal scale to another. We now describe some of these methods.

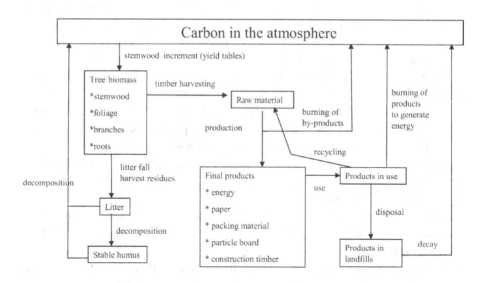

Figure 5.3. Components of a full forest sector carbon balance based on forest inventories. These all together provide a full forest sector carbon balance. This method is clearly converging across studies (Birdsey, 1992; Kurz and Apps, 1999; Karjalainen, 1996; Karjalainen *et al.*, (in press); Nabuurs *et al.*, 1997; Nilsson *et al.*, 2000).

3.   Forest inventories

Figure 5.3 displays a set of estimates that have been based on forest inventories. Full carbon budgeting based on forest inventories relies to a great extent on representative, long term measurement series of stemwood volume and increment. Forest inventories are traditionally carried out to inform forest managers about the state of their forests in terms of area, species, age classes, growing stock (quantity of wood), net annual increment, and fellings. All of these measurements are typically carried out on the stemwood only, and have traditionally been carried out by ground measurements, but are increasingly a combination of remotely sensed data and ground data.

A forest inventory is usually carried out on a network of sample plots. The design and intensity of the plots depends on the forest heterogeneity, and costs. For example, 10 European countries have forest inventories in which a single field plot (usually 25 trees) represents around 200 to 1000 ha (Köhl and Païvinen, 1997). The total number of measurement plots thus amounts to 424,000. In the United States national forest inventory field plots represent a range from 2000 ha to 4700 ha depending on factors such as growth rates, frequency of disturbance, and accessibility. Inventories are usually carried out in cycles of 5–10 years. This means that the annual variability in growth rates is captured in a less accurate way. Inventories can be carried out in such a way that they yield very accurate results; for example at the country level producing uncertainties (95% confidence) for forest land area of ±0.4%, growing stock 0.7%, and total increment ±1.1% (Tomppo, 1996).

Continuous forest inventories have been carried out for all Annex B countries of the Kyoto Protocol, and in some cases as far back as the 1920s. These data cover the complete northern hemispheric temperate and boreal forests. However, between these countries there are differences in precision and definitions. Variables reported by one country can therefore not always be compared to the same variable for another country (Köhl and Païvinen, 1997). The TBFRA2000 project (UN-ECE/FAO, 2000) seeks to encourage data gathering based on a harmonised set of definitions.

The enormous set of data coming from forest inventories provides a unique opportunity for assessing a full forest sector carboncarbon balance (see Figure 5.4). Even though the stemwood and the fellings are the only directly measured data, methods exist to scale this up to a full forest sector carbon balance. The other tree components are added based on conversion coefficients, and with turnover rates at which they produce litter. For the litter and soil compartments, usually a modelling tool is used. The harvesting and wood products compartments is usually modelled using felling statistics and life span estimates. These all together provide a full forest sector carbon balance; either as a static conversion for a base year, or by using a dynamic modelling tool for providing projections under alternative management regimes. However, it is clear that with adding all these components, the uncertainty increases as well.

The inventory based carbon budgeting methods have their limitations. While the assessments of stemwood volume are very accurate, adding many components increases the uncertainty. Inventories quantify the effects of diverse mechanisms, but they yield limited information about the importance of individual mechanisms. Furthermore they are carried out over cycles of 5–10 years and thus do not cover the annual variability very well, and inventory methods do differ in details between countries. Still the inventory based carbon budgeting methods are converging towards each other, and provide a unique methodology that builds upon an accurate dataset consisting of billions of individual tree measurements.

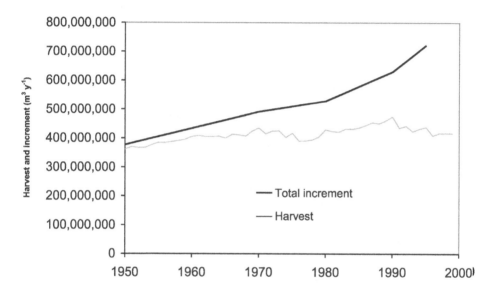

Figure 5.4. Development of total European forest increment in terms of stemwood and total European harvest of wood since 1950.

The figure shows that the increment has increased steadily since 1950, while the harvest has stayed approximately the same (based on data from UN-ECE/ FAO). The difference is the build-up of biomass in the forest, and indirectly shows an increasing sink in the forest biomass of European forests. This sink can probably continue for decades, but will eventually saturate and become a source.

4. Flux measurements

The European flux monitoring system set up as EUROFLUX (Valentini *et al.*, 2000) has shown that it is possible to continuously monitor the net exchange of carbon by terrestrial ecosystems. Perhaps more importantly, this technique allows us to verify the net carbon balance, which is required by the Kyoto protocol. Martin *et al.* (1998) estimated using forest maps and yearly Net Ecosystem Exchange (NEE is equivalent to NEP) data from Euroflux, that between 0.17 and 0.31 Gt C was sequestered by European forest in 1997 (Figure 5.5). The flux estimate ignores harvest and any wood taken out of the ecosystem and depends on, not always accurate, information on forest distribution within Europe. Furthermore, research sites are typically chosen in well-maintained forest without major disturbances, so the average estimate of NEE of 3 ton C ha^{-1} yr^{-1} is almost certainly an upper limit: one could almost call this 'potential NEE'.

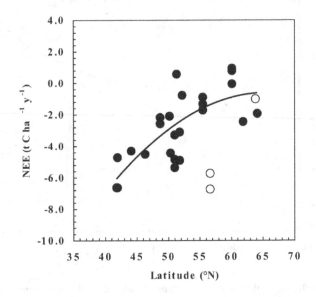

Figure 5.5. Variation of Net Ecosystem Exchange of European forest with latitude (redrawn after Valentini *et al.*, 2000). The open circles represent plantation forests in warm, maritime climates.

Figure 5.5 shows results obtained from the European flux network and the variation of NEE with latitude. The interpretation of this graph generated considerable controversy after its publication in Valentini *et al.* (2000), as seen for instance in Piovesan and Adams (2000) and Jarvis *et al.* (2001). Surprisingly, the graph shows that forests at the Mediterranean belt of the temperate zone in Europe, have the largest uptake (highest absolute value of NEE). This is counterintuitive, as forest inventories show the largest growth in Northern, Scandinavian areas. However as shown in Figure 5.2, the NEE is the result of two large fluxes, and what the graph in fact hides is the little variation in photosynthetic uptake but the large variation in respiratory fluxes as found in Valentini *et al.* (2000). In fact, the Italian forests have an annual respiration of about 5 tons C per hectare, while the more northern forest have values of almost 10 tons ha^{-1} yr^{-1}.

But, despite the controversy over the precise interpretation of this plot and over the question why there appears to be no clear relation of respiration with mean annual temperature (Giardina and Ryan, 2000), there are a number of conclusions to be drawn from the information obtained from eddy correlation networks. The variation in NEE is not dependent on annual mean values of temperature, rather short term processes like onset of thawing in boreal forest (Lindroth *et al.*, 1998) and radiation loading in the summer in mid-latitude forest (Dolman *et al.*, 2000) appear to drive the variation in NEE. It is likely that the variation in NEE with latitude is caused by a mix of factors, among which the length of the growing season may play a dominant role. The eddy correlation network also generates rather precise information over the main factors driving NEE at short time scales and as such are useful in improving process level models (Kramer *et al.*, 2001).

5. Inversion based estimates

The atmospheric inversion method results in values for CO_2 uptake, not just biospheric uptake, and depends heavily on accurate estimates of fossil fuel emission, to obtain the sink strength as a residual in the optimisation process. The atmospheric CO_2 concentrations from a number of measurement sites constrain the source/sink areas in the presence of estimated atmospheric transport and known other sources and sinks. It is the latter aspect, or the initial condition of the modelling, which determines to a large extent the outcome.

Since the typical atmospheric gradient of CO_2 from North to South Pole is only 3 ppm, the accuracy of the network is critical. Up to now, only the NOAA network provides modellers with data of the required accuracy at the global scale. From Figure 5.6 it can be seen that most of the flask sampling locations are based on islands in the ocean. Whilst this may be the best location to obtain pollution free samples of air with so-called tropospheric CO_2 values, this is not necessarily the best location to estimate the terrestrial uptake of CO_2.

The latitudinal gradient in observed CO_2 concentration indicates a northern hemispheric sink. Furthermore O_2/N_2 and stable isotope measurements confirm a location in the Northern Hemisphere. At this moment in time there is no hard scientific evidence to either favour a Northern US or Eurasian sink above a tropical one. The high estimate of Fan et al. (1998) for the US is almost certainly the result of a small sampling in time and thus represents an atypical situation.

Current inversion models can distinguish at best broad latitudinal bands. That approach locates the terrestrial sink, estimated as -0.6 to -2.2 Gton yr^{-1}, primarily in the Northern latitudes (>300), with the uptake in the tropics balanc-

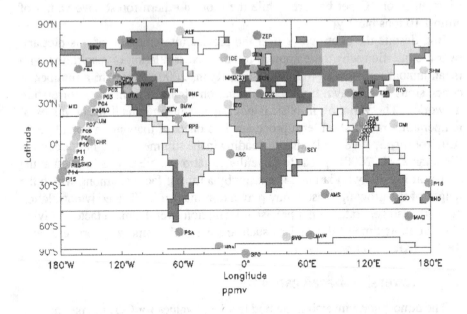

Figure 5.6. Locations of the flask network of NOAA plus additional sites used in the Bousquet *et al.* (2000) inversion model study.

ing the deforestation (1.7 Gton yr^{-1}). Bottom up approaches using terrestrial models are not always in agreement with this location and sometimes suggests more uptake in the tropics than in the extratropics. There is still a considerable imbalance between the estimates obtained by inverse models and other studies, with the modelling studies generally providing the lower bounds and the inverse models the upper bounds. Going into more regional detail is virtually impossible, as shown by the current best estimates from various methods 0.2–1.3 Gt yr^{-1}, conterminous US 0.2–1.3 Gt yr^{-1} and Siberia 0.01–1.3 Gt yr^{-1} and for Europe 0.08–0.4 Gt yr^{-1}. More recent unpublished work from forest inventories for Europe suggests a forest sink of 0.086 Gt C yr^{-1} for the period of 1980–1999. During 1950 to 1970 the sink was 0.023 Gt C yr^{-1}, which is only 10% of the best current inversion based estimate.

6. Forward modelling

Many Terrestrial Ecosystem Models (TEMs) and Dynamic Global Vegetation Models (DGVMs) suggest that the global land carbon sink may saturate sometime during this century (Cramer *et al.*, 1999). These models describe the main biogeochemical processes and treat the elementary cycles of carbon, nitrogen, and water. They vary widely in the level of detail of these descriptions. On the whole these models produce an increase in land carbon storage as a result of rising CO_2 but a reduction as a result of the associated climate warming (which enhances soil and plant respiration). The balance between these two competing effects is predicted to change as CO_2-fertilisation of photosynthesis saturates at high CO_2, but respiration is assumed to increase with temperature. As a result the current global land carbon sink is projected to increase less and less rapidly, with some models even predicting its conversion to a source. The wide range of projections from TEMs results from basic differences in the way key processes are modelled, and these differences themselves highlight gaps in basic process understanding (Figure 5.8). For example the standard assumption that the rate of soil respiration increases significantly with temperature has recently been challenged by observational evidence (Giardina and Ryan, 2000). Such uncertainties project on to predictions of climate change as well as land carbon uptake.

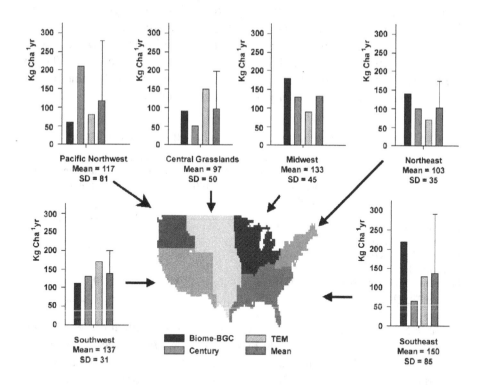

Figure 5.7. Net carbon storage for the US. Estimated with three terrestrial ecosystem models. Histograms denote specific model results and the mean (95% confidence interval) (redrawn after Schimel *et al.*, 2000).

Schimel *et al.* (2000) compared the performance of three bottom up terrestrial ecosystem models for the conterminous US and found that model differences in net carbon storage by ecosystems were large. In estimating the continental mean carbon storage the agreement was within 25%. The real advantage of using the bottom up class of models is that potentially the causes for the observed large sink in recent years may be distinguished. These causes include climate variation, increased nitrogen deposition and the CO_2 fertilisation effect. For the US, the VEMAP project found that for a sink of 0.3 Gt C yr^{-1}, the effect of CO_2 fertilisation and climate amounted to 0.1 Gt C yr^{-1}, while the remaining sink of 0.2 Gt C yr^{-1} was attributed to re-growth on abandoned land and harvested forest land. Schimel *et al.* (2000) claim that land use history and management are far more important factors in determining the current sink strength than climate and CO_2 fertilisation effects. This is in broad agreement with the results obtained in the forest inventory studies in Europe.

Recent work using the first General Circulation Models (GCMs) to include the carbon cycle interactively (with a DGVM), suggest that climate change will reduce the ability of the land biosphere to absorb emissions (Cox *et al.*, 2000; Friedlingstein *et al.*, 2001). However, the magnitude of this effect varies between the models, ranging from a slight reduction (Friedlingstein *et al.*, 2001), to a climate-driven conversion of the global land sink to a source by 2050 (Cox *et al.*, 2000). If confirmed, the latter result would seriously undermine the use of land sinks as a long-term alternative to cutting emissions.

Table 5.1. Global warming and the terrestrial carbon cycle (from Cox *et al.*, 2000)

Experiment	Mean surface warming over land	CO_2 concentration in 2100 (ppm)	C-uptake in 2000–2100 (Gton C)	
			Land	*Ocean*
Predefined CO_2 emissions without global warming	0.0°C (0.0°C)	700	450	300
Predefined CO_2 concentration with global warming	4.0°C (5.5°C)	713	−60	250
Predefined CO_2 emissions with global warming	5.5°C (8.0°C)	980	−170	400

Table 5.1 shows results of the first coupled climate carbon cycle study executed by *Cox et al.* (2000), while Figure 5.9 shows the projected changes in carbon stocks of global soils and vegetation. The time evolution of the stocks of carbon is illuminating. Until 2050 the land surface continues to act like a sink and stores carbon. After that it turns into a source, because large areas of tropical rainforest die off and release carbon, and respiration overtakes photosynthesis as the dominant process in NEE because of temperature increases. Consequently the CO_2 concentration in the fully coupled run is higher than in a run where the biosphere was not allowed to emit CO_2 back into the atmosphere. Table 5.1 shows a comparison of several runs with this coupled system. Their

run 1 (predefined CO_2 emissions without global warming) is a standard run with a GCM where no feedback of the increase of CO_2 on the resulting climate is allowed. Such a simulation suggests that the land takes up 450 and the ocean 350 G C from 2000 to 2100. Allowing the climate to respond to the increase in fossil fuel CO_2 only dramatically alters this picture, now the land acts as a source and the ocean sink is somewhat smaller. The overall climate effect of this is a 5.5°C warming over land. If the additional CO_2 produced by the land is allowed to go back into the atmosphere the warming increases even further (8°C over land) and the land surface produces even more CO_2 due to large scale dying of the Amazonian rainforests and increasing soil respiration (Figure 5.9).

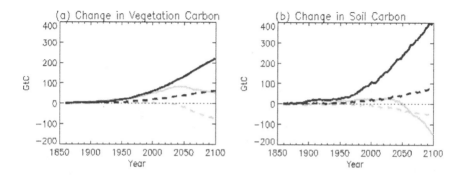

Figure 5.8. Effect of global warming on changes in land carbon storage. The gray lines represent a fully coupled climate/carbon-cycle simulation, and the black lines are from an 'offline' simulation which neglects direct CO2-induced climate change. The figure shows simulated changes in vegetation carbon (a) and soil carbon (b) for the global land area (continuous lines) and South America alone (dashed lines) (Redrawn from Cox *et al.*, 2000).

7.   Interannual variability

The uptake of carbon by the terrestrial biosphere was lower in the eighties than in the nineties of the previous century (IPCC, 2000). It is unclear what the precise cause for this enhanced uptake of carbon is. One inversion model result suggests that in the 1980s the tropics contributed to most of the observed interannual variability, whereas in the 1990s the northern hemispheric lands were the main source of interannual variability. It is believed that a considerable part of the global interannual variation in the CO_2 concentration can be attributed to biomass burning in the tropics. Land fluxes are almost twice as variable as the ocean fluxes (Bousquet *et al.*, 2000). Variation in land uptake can be of the same order as the mean value. Measurements from inversion models show an annual global variation of 5 Gt C yr^{-1}.

Measurements from flux tower sites show a variation of 50% of the NEE (variation of 1–2 t C ha^{-1}). As an example, the annual carbon balance of a mid-latitude pine forest in the Netherlands is shown in Figure 5.10. The highest NEE is obtained in 1996 with a value of 4.42 t C ha^{-1}. The lowest uptake took place in 1998 (2.76 t C ha^{-1}). At the annual time scale the variation in assimilation is again larger than the variation in respiration. There is no relation between annual mean values of respiration and temperature. Also, at the annual time scale there is no relation between assimilation and radiation. Interannual variability is primarily driven by processes operating at time scales smaller than a year.

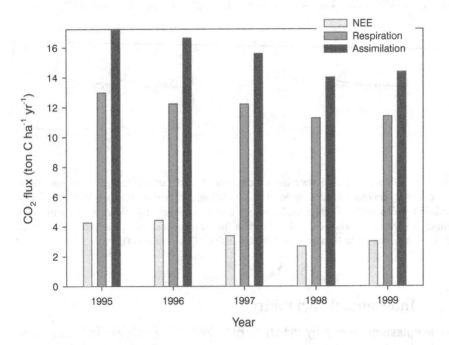

Figure 5.9. NEE, assimilation and respiration for a Dutch Scots Pine forest of 80 years age, obtained by eddy correlation and gap filled with neural network techniques.

Inversion model based flux anomalies over Western Europe are on the order of ∼0.5 Gt C over the past 20 years. In 1996, Europe was an anomalous CO_2 source of 0.6 Gt C. Anomalies in the western European carbon balance appear to correlate with North Atlantic Oscillation changes, the oscillation that drives a considerable part of the West European weather. The negative phase of the North Atlantic oscillation, with low NAO (North Atlantic Oscillation index) years associated with wintertime conditions being dryer and colder over Northern Europe and wetter over the Mediterranean and appears to be synchronous with an anomalous carbon release by European ecosystems (Figure 5.10). The converse appears to hold for the positive NAO being associated with anomalous carbon uptake. These correlations are stronger for the southern hemisphere Enhanced Southern Oscillation (ENSO). However, these major variations in global climate and associated CO_2 uptake provide a strong check on the performance of coupled models (e.g., Cox *et al.*, 2000).

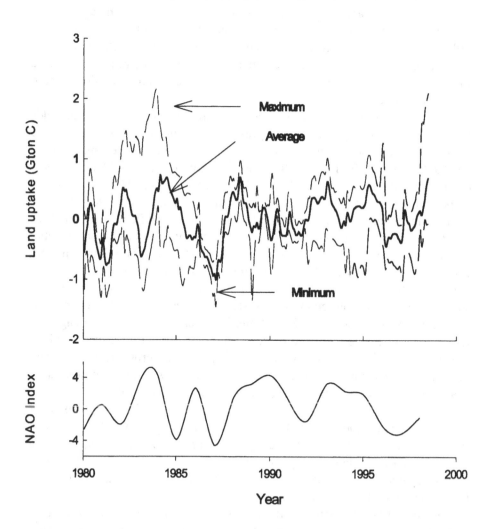

Figure 5.10. Inferred land uptake for West Europe and NAO index. Uptake is derived from latest inverse model results (e.g., Bousquet *et al.*, 2000 and Valentini *et al.*, 2000).

8.　　CO$_2$ uptake of non-forest and agricultural land

Most studies on CO$_2$ uptake, except perhaps the Schimel *et al.* (2000) study, have focused on forested ecosytems because they are the most obvious long-term absorbers of carbon. Agriculture is usually treated in global models as carbon neutral. However this assumption is hardly based on strong scientific evidence and some attention to the carbon use of agricultural or non-forest systems is therefore justified.

By far the most important carbon stock in both grasslands and arable lands is the soil organic carbon (SOC). According to IPCC (2000) many agricultural soils are currently sources of carbon both in tropical and temperate zones as soil carbon continues to decline. Changes in agricultural practices for other than climate policies tend to mitigate this trend. This would reduce the source strength of agricultureagriculture and could reduce carbon emission or turn agricultural soils into a sink.

Terrestrial ecosystems have significant potentials to store carbon. This holds for many if not all ecosystems (forest, grassland, pasture, agricultural land). The IPCC's Second Assessment Report (SAR) estimated the storage capacity until 2050 at 60–87 Gt C for forests (Brown et al., 1996) and at 23–44 Gt C for agricultureagriculture (Cole *et al.*, 1996). This adds up to 90–130 Gt and is 1.8–2.6 Gt per year over 50 years. In the SAR estimates, not all biomes, activities and pools were included. Sampson *et al.* (2000) in the Special Report on Land Use, Land Use Change and Forestry estimated the potential sequestration potential of additional activities at 1.3–2.5 Gt C per year in the first commitment period. This estimate in the SRLUC includes more C pools and more biomes that did the SAR estimate and involves new activities started to explicitly increase C stores in natural and agricultural ecosystems. So, the SRLULUC estimate is more conservative than the SAR estimate. In general, carbon stocks in agricultureagriculture can be increased by adjusting management for a specified land use (no tillage, fertilisation) or by changing land use (cropland to grassland to natural ecosystems). Significant stocks of carbon exist below ground; their exact size is less well known that that of the above ground pools.

Sampson *et al.* (2000) is the most recent and comprehensive assessment of the potential to store carbon through management of terrestrial ecosystems. Sampson *et al.* (2000) in the SRLUC gives estimates on carbon sequestration for cropland management in dry temperate zones of 0.1–0.3 tC ha^{-1} yr^{-1} and in wet temperate zones of 0.2–0.4 tC ha^{-1} yr^{-1} and grassland management of 0–0.3 and 0.4–2.0 tC ha^{-1} yr^{-1} respectively.

It is possible to develop a model that reflects the inputs of carbon in agricultural systems and that, at the same time, is constrained by actual production data of the FAO (Vleeshouwers and Verhagen, 2002).

This model, CESAR (Carbon Emission and Sequestration by AgRicultural land use) keeps track of the SOC by calculating the input and output of carbon to or from the SOC. The model was applied to Europe, with a special focus on the Netherlands. For the Netherlands, data from LEI/CBS were used whereas for Europe the FAO data-set was used (FAO, 2000). The model evaluates the separate contribution of variation in weather conditions, rising temperature, rising atmospheric CO_2 concentration, increase in agricultural production over the years, and specific measures such as reduced tillage, application of organic or green manure, or leaving crop residues in the field.

Annual data on crop production levels are readily available (FAO, CBS/LEI). Given a ratio between harvested and non-harvested products these production levels can be used to determine the amount of crop residues and the possible carbon input when leaving the residues on the field. Based on FAO production data for 1998, simulations with CESAR were made to estimate the effects of the above-mentioned measures and developments.

The results of these were calculated for arable fields in all EU countries, and are shown in Table 5.2. FAO production data for 1998 were used, while the starting date of the measures was 2000. The initial soil carbon content was set at $80\,t\,ha^{-1}$, agreement with the value for arable fields given by IPCC (2000). For each country the area of the six major crop groups (Cereals, Roots and Tubers, Sugar beets, Pulses, Oilcrops, Fibre Crops, Vegetables and Melons) was used to derive a country-specific crop mix. Long-term average weather data used were obtained from FAO (1990).

Table 5.2. Average predicted annual net effect on SOC in arable fields in the commitment period ($tC\,ha^{-1}\,yr^{-1}$) resulting from different treatment.

Country	CO_2 increase	Straw left	Farmyard manure	Green manure	Temp. increase	Reduced tillage	Conversion to meadow
Austria	0.011	0.2	1.4	0.4	−0.01	0.23	2.7
Belg & Lux	0.014	0.3	1.4	0.5	−0.04	0.25	2.2
Denmark	0.012	0.3	1.4	0.4	−0.02	0.22	3.8
Finland	0.007	0.2	1.5	0.2	−0.02	0.16	2.4
France	0.013	0.3	1.5	0.5	−0.02	0.20	2.1
Germany	0.012	0.3	1.4	0.4	−0.03	0.22	3.1
Greece	0.005	0.1	1.5	0.3	−0.01	0.16	0.9
Ireland	0.012	0.3	1.4	0.4	−0.04	0.24	2.6
Italy	0.010	0.2	1.5	0.4	−0.03	0.20	2.5
Netherlands	0.013	0.1	1.4	0.5	−0.04	0.24	2.6
Portugal	0.004	0.1	1.5	0.2	−0.03	0.18	0.9
Spain	0.006	0.1	1.6	0.3	−0.01	0.13	3.3
Sweden	0.010	0.3	1.5	0.3	−0.02	0.17	1.4
UK	0.013	0.3	1.4	0.5	−0.03	0.21	2.4

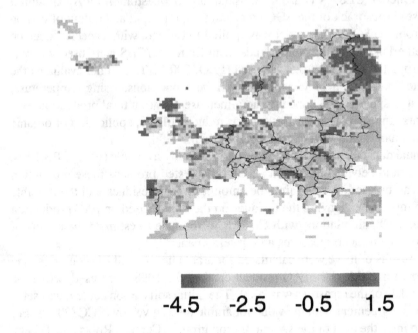

-4.5   -2.5   -0.5   1.5

Figure 5.11.   The net effect of Business as Usual (BAU) for Europe in the commitment period 2008–2012 in ton C ha^{-1} sequestered carbon.

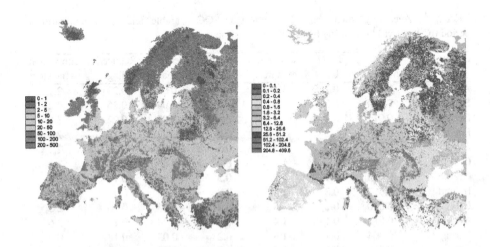

Figure 5.12.   Above (a) and below ground (b) carbon stocks based on forest inventory and FAO production data, linked to a land use classification map derived from NOAA AVHRR and the FAO soils database.

The results presented here are preliminary and aim to show the potential of the model. Several aspects of CESAR need to be tested with more independent data series. In particular the effect of low soil moisture contents on the decomposition of SOC will need to be validated. In the present version the annual decomposition rate of SOC is lowest in Mediterranean countries because of summer drought and in northern countries because of low temperatures. Estimates made by the model may also be improved by the availability of data on the amount of crop residues produced in different countries, and on the production in grasslands. The results of Table 5.2 suggest that the net carbon uptake or loss is relatively independent of the initial carbon content in the soil. Calculation of the absolute changes in SOC, however, strongly depends on the initial carbon content.

When the average annual carbon flux in the EU during the commitment period 2008–2012 is plotted as a function of the initial carbon content in 2000 a linear relation is obtained. When the present amount of SOC is below 38 t ha^{-1}, arable fields are sinks of carbon, whereas above 38 tC ha^{-1}, arable fields are sources of carbon. For meadows the steady state value is 230 tC ha^{-1}. Estimates given in the IPCC Special Report are 80 tC ha^{-1} for arable fields and 236 tC ha^{-1} for temperate grassland. These values would imply that arable soils in the EU would be a carbon source amounting to 31 MtC yr^{-1}, and grasslands a carbon source of 5 MtC yr^{-1} during the Kyoto commitment period. It is worth emphasising that the actual carbon content of agricultural soils is poorly known for Europe, let alone for less populated parts of the world. Improvement of these estimates would therefore primarily depend on improved accuracy of the soil organic matter content.

Taking the global soil data as provided by the IGBP-DIS to establish the initial condition of SOC and the average climate conditions as provided by the Climatic Research Unit in Norwich, UK (www.cru.uea.ac.uk) the Business as Usual scenario can be calculated in a spatially explicit manner.

The net effect of BAU for arable farming, as calculated with CESAR, for the 2008–2012 commitment period is given in Figure 5.11. Masking all non-arable land uses will allow a more detailed overview. Both temperature and moisture effects are accounted for in the calculations for Figure 5.11.

9. A preliminary Carbon Inventory for Europe

One of the key questions raised by the Kyoto protocol is how to calculate the changes in 'carbon stocks' associated with land use changes and forestry activities during the commitment period. 2008–2012. There is no up-to-date generally accepted database of the sink strength of for instance the European landscape. This hinders current attempts to precisely locate the main sources and sinks.

To provide a first estimate of the carbon sequestration potential of European land, a 'quick and dirty' inventory was made using an up-to-date land-cover classification map derived from remote sensing. Carbon sequestration potential was allocated through a geographic information system to the various land covers. Each land-cover class thus has its own C-uptake. For forests this was derived from national inventories, for agricultural land simple models and FAO production data were used. For the soil carbon content the FAO (IGBP-DIS) soil map was used. This procedure resulted in estimates of above and below ground carbon content as shown in Figure 5.12. What is evident from this figure is that the carbon content of the vegetation appears to be dependent of the political boundaries of the countries. This is caused by the fact that national inventories differ between countries. On the whole however a reasonable pattern of carbon content over Europe emerged. Such data based inventories can be used to check submissions by individual countries, but also as a means to track changes over time. They may also be used to serve as a bridge between the site-based estimates and continental scale inversion-based estimates.

10.   Discussion

One of the key questions relates to the future of the carbon sink: will it saturate, and if so, when? To predict when sink saturation may occur, we need to understand which are the main limiting factors to carbon uptake. Apart from management and disturbance, the main factors controlling primary productivity in sinks are atmospheric CO_2 concentrations, nutrient availability, water availability, and absorbed solar radiation. Usually these factors interact, where one of them is of overriding importance, depending on location and climate. In temperate and boreal regions, nitrogen is often this main limiting factor, although this role is quickly taken over by atmospheric CO_2 concentrations. Recent studies have shown only limited dependence on nitrogen availability and nitrogen fertilisation, but if increases in N are combined with increases in CO_2, the effect may be much larger than the effects of either of these factors alone.

The pool of organic soil carbon serves as an ecological memory that documents human influence and natural disturbances for decades, even centuries. Depending on several environmental conditions, i.e., climate, soil composition and management, the respiratory losses or ecosystem assimilation component of the C-cycle may dominate, causing in the first case, the system to act as source, and in the second as sink. It has long been known that the rate at which soil carbon is being turned over depends also on nitrogen content, thus fuelling the hypothesis that nitrogen fertilisation is the main cause for the accelerated growth rate of forests. Recent studies however, (Schulze, 2000) over a South to North transect in Europe indicate that anthropogenic N did not influence the C and N turnover rates significantly, at the short term. A similar finding was recently reported by Nadelhoffer *et al.* (1999) who administered labelled nitrogen to nine forests and found that little of the nitrogen administered was found back in the biomass. Thus, high levels of primary production are likely not to be caused by the nitrogen fertilisation. Trees do not appear to be able to use the nitrogen present.

Several lines of theory and evidence predict that through nitrogen fixing organisms, ecosystems will optimise their nitrogen content according to the available radiation and soil water – in which case enhanced N deposition will not result in great changes in uptake. This is likely to be the case in tropical rain forests, and recent studies showed a significant potential effect of further atmospheric CO_2 increase in these regions. In tropical soils such as in the Amazon, the main other limitation is likely to be low phosphorus availability. However, it has been suggested that also this limitation might be 'broken' since accumulation of complex organic compounds in these soils will alter the chemical balance in such a way that progressive amounts of phosphorus will be mobilised from otherwise inaccessibly bound soil complexes. Most of these mechanisms are as yet only the result of theoretical and modelling studies, let alone predictions of possible future sink saturation.

Decomposition of organic material in litter and soils, in which process CO_2 is released back to the atmosphere, is done by micro-organisms, whose enzyme activities are essentially temperature dependent. Direct measurements of CO_2 emissions from soils from single sites, done over limited time spans, usually also show a clear dependence on soil temperature. This has led to the suggestion in several studies that current carbon sinks in ecosystems may be extremely vulnerable to climate change, and likely to change into sources in the future, since the processes governing emission are much more temperature-dependent than those controlling uptake. This has led to sometimes dramatic 'doom' scenarios in which whole forest areas decline progressively in a feed-forward process of carbon losses to the atmosphere and increased warming.

However, several recent studies (e.g., Giardina and Ryan, 2000) show that emission fluxes, when compared between ecosystems and over longer time intervals, are less strongly related to temperature but much more to the amount of organic material and the input of organic matter through NPP and therefore indirectly to the absorbed radiation. The consequences of this for, for instance, the coupled GCM DVM studies need to be further explored.

At COP VI decisions were to be made as to what land use and forestry activities can be counted as emissions reductions. The inclusion of sinks in the Kyoto protocol would then be a fact, underlining the important role, the terrestrial biosphere plays in the global carbon balance. Our current understanding of the role of the biosphere in the global carbon balance however may not always be sufficient to adequately underpin these decisions. For some, realistic monitoring methodologies can be agreed as shown above, for some, the potential and actual carbon sequestration results may be still unknown and hard to determine. A cautious approach whereby estimates of the scientific uncertainty are used to determine what activities to include and what activities not, may be the most appropriate approach. The inclusion of sinks in the Kyoto protocol is a major step forward in realising a comprehensive carbon accounting system of the earth, but we continue to need to improve our understanding of the terrestrial biosphere to make the Kyoto protocol really work. The fact that the negotiations at the Hague stalled on the sinks issue is an indication that policy makers are beginning to be aware of this.

Acknowledgments

A considerable amount of the work described in this paper was carried out in the context of the NRP (Dutch National Research Program on Climate Change and Global Air Pollution) funded projects: Post Kyoto accounting of the carbon balance of Europe (no 958254) and The scientific basis behind key decisions on sinks and the CarboEurope projects funded by the Fifth Framework Program of the European Union Region Assessment of the Carbon Balance of Europe (EVK-2199-00236). Most of the writing of this chapter was done while the first author was employed at Alterra, Wageningen-UR.

References

Birdsey, R.A. (1992) Carbon storage and accumulation in U.S. forest ecosystems, USDA Forest Service General Technical Report WO-59

Bousquet, P., Peylin, P., Ciais, P., le Quere, C., Friedlingstein, P and Tans, P. (2000) Regional changes in carbon dioxide fluxes of land and oceans since 1980. Science 290, 1342–1346.

Brown, S., Sathaye, J., Cannell, M., and Kauppi, P. (1996) Management of forests for mitigation of greenhouse gas emissions (Chapter 24). In: R.T. Watson, M.C. Zinyowera, R.H. Moss and D.J. Dokken. (Eds.), Climate Change 1995. Contribution of Working Group II. Cambridge University Press, Cambridge, UK, 773–797.

Ciais, P., Tan, P.P., Trolier, M., White, J.W.C. and Francy, R.J. (1995) A large northern hemisphere terrestrial CO_2 sink indicated by 13C/12C of atmospheric CO_2. Science 269, 1098–1102.

Cole, C.V., Duxbury, J., Freney, J., Heinemeyer, O., Minami, K., Mosier, A., Paustian, K., Rosenberg, N., Sampson, N., Sauerbeck, D. and Zhao, Q. (1996) Agricultural options for mitigation of greenhouse gas emissions. In: Watson, R.T., Zinyowera, M.C. Moss R.H. and Dokken. D.J. (eds.) Climate Change 1995. Impacts, Adaptations, and Mitigation of Climate Change: Scientific-Technical Analysis. Cambridge University Press, 745–771.

Cox, P.M., Betts, R.A., Jones, C.D., Spall, S.A. and Totterdell, I.J. (2000) Acceleration of global warming due to carbon cycle feedbacks in a coupled climate model. Nature 408, 184–187.

Cramer, W., *et al.* (1999) Comparing global models of terrestrial net primary productivity (NPP): overview and key results. Glob. Change. Biol., 5, 1–15.

Dolman, A., Moors, E.J., Elbers, J.A. and Kruijt, B. (2000) Inter-annual variability in co2 exchange of a mid latitude temperate pine forest. Proceedings of the first ASIA-Flux conference. Sapporo.

Fan, S., Gloor, M., Mahlman, J., Pacala, S, Sarmiento, J. Takahashi, T. and Tans, P. (1998) A large terrestrial carbon sink in North America implied by atmospheric and oceanic carbon dioxide data and model. Science 282, 442–446.

FAO (2000) FAOSTAT Agricultural Database. Internet http://www.fao.org

FAO (1990) FAOCLIM Global Weather Database. Agrometeorology group, Remote Sensing Centre.

Friedlingstein, P, Bopp, L., Ciais, P., Dufresne, J-L., Fairhead, L., LeTreut, H., Monfray, P and Orr, J. (2001) Positive feedback between future climate change and the carbon cycle. Geophys. Res. Letters, 28: (8), 1543–1546.

Giardina, C.P and Ryan, M.G. (2000) Evidence that decomposition rates of organic carbon in mineral soil do not vary with temperature. Nature 404, 858–861.

Houghton, R.A., Hackler, J.L., Lawrence, K.T. (1999) The US carbon budget: contributions from land use change. Science 285, 574.

IPCC Second Assessment Report (1996) Climate Change 1995: Impacts, Adaptations and Mitigation of Climate Change: Scientific-Technical Analyses. Contribution of working group II to the Second Assessment Report of the Intergovernmental Panel on Climate Change. University Press, Cambridge, UK.

IPCC (2000) Land Use, Land Use Change and Forestry. R.T. Watson, I.R. Noble, B. Bolin, N.H. Ravindranath, D.J. Verardo and D.J. Dokken (eds.). A Special report of the IPCC. Cambridge University Press, Cambridge, UK.

Jarvis, P.G., Dolman, A.J., Schulze, E.-D., Matteucci, G., Kowalski, A.S., Ceulemans,R, Rebmann, C., Moors, E.J., Granier. A., Gross, P., Jensen, N.O., Pilegaard, K., Lindroth, A., Grelle, A., Bernhofer, Ch., Grnwald, T., Aubinet, M., Vesala, T., Rannik, ., Berbigier, P., Loustau, D., Gudmundsson, J., Ibrom, A., Morgenstern, K., Clement, R., Moncrieff, J., Montagnani, L., Minerbi, S. and Valentini, R. (2001) Carbon balance gradient in European forests: should we doubt the surprising results? A reply to Piovesan and Adaman. J. Veget. Sci. 12, 145–150.

Kaminski, T, Heimann, M., Giering, R. (1999). A coarse grid three dimensional global inverse model of the atmospheric transport. 1. Adjoint model and Jacobian matrix. J Geophys. Res. 104, 18535–18553.

Karjalainen, T. (1996) Dynamics of the carbon flow through forest ecosystem and the potential of carbon sequestration in forests and wood products in Finland. Research Notes Faculty of Forestry, University of Joensuu, Finland Academic dissertation.

Karjalainen, T., Liski, J., Nabuurs, G.J. and Pussinen, A. (In press.) An approach towards an estimate of the impact of forest management and climate change on the European forest sector carbon budget Presentation held at Potsdam, NIMA conference.

Kauppi, P.E., and Tomppo, E (1993) Impact of forests on net national emissions of carbon dioxide in West Europe. Water Air and Soil Pollution 70, 187–196.

Kauppi, P.E., Mielikainen, K. and Kuusela, K (1992) Biomass and carbon budget of European forests, 1971 to 1990. Science 256, 70–74.

Köhl, M. and Païvinen, R. (1997) European Union, Study on European forestry information and communication systems, vol. 1& 2, Brussels, Belgium.

Kurz, W.A., and Apps, M.J. (1999) A 70-year retrospective analysis of carbon fluxes in the Canadian forest sector, Ecological Applications, 9(2), 526–547

Kramer, K., Leinonen, I., Bartelink, H.H., Cienciala, E., Froer, O., Gracia, C.A., Hari, P., Kellomki, S., Loustau, D., Magnani, F., Matteucci, G., Nissinen, A., Sabat, S., Sanchez, A., Sonntag, M., Berbigier, P., Bernhofer, C., Dolman, A.J. Moors, E.J. Jans, W.A. Granier, A., Grnwald, T., Valentini, R. Vesala, T. and Mohren, G.M.J. Evaluation of 6 process-based forest growth models based on eddy-coviance measurements of CO_2 and H_2O fluxes at 6 forest sites in Europe. Global Change Biol. in press

Lindroth, A., Grelle, A., and Moren, A.S. (1998) Long term measurements of boreal forest carbon balance reveal large temperature sensitivity. Global Change Biol. 4, 443–450.

Martin P., Valentini, R., Jaqcues, M., Fabbri, K., Galati, D., Quarantino, R., Moncrief, J.B., Jarvis, P., Jensen, NO., Lindroth, A., Grelle, A., Aubinet, M., Celemans, R., Kowalski, A.S., Vesala, T., Keronen, P., matteuci, G., Granier, A., Berbingier, P., Lousteau, D., Schulze, E.D., Tenhunen, J., Rebman, C., Dolman, A.J., Elbers, J.E., Bernhofer, C., Grunwald, T. and Thorgeirson, H. ., 1998. New estimate of the Carbon sink strength of EU forests integrating flux measurements, field surveys and space observations. Ambio 27, 582–584.

Nabuurs G.J., Païvinen, R. Sikkema, R., and Mohren, G.M.J. (1997) The role of European forests in the global carbon cycle – a review, Biomass and Bioenergy, 13(6), 345–358.

Nadelhoffer, K.J., Emmett, B.A., Gundersen, P., Kjnaas, O.J., Koopmans, C.J., Schleppi, P., Tietema, A., and Wright, R.F. (1999) Nitrogen deposition makes a minor contribution to carbon sequestration in temperate forests. Nature 398, 145–148.

Nilsson, S., Shvidenko, A., Stolbovoi, V., Gluck, M., Jonas, M. and Obersteiner, M. (2000) Full carbon account for Russia IIASA Interim Report IR-00-021. Laxenburg, Austria.

Piovesan, G. and Adams, J.M. (2000) Carbon balance gradient in European forests: intepreting EUROFLUX. J. Veg. Sci. 11, 923–926.

Rayner, P.J. and Law, R.M. (1999) The internannual variability of the global carbon cycle. Tellus Ser B 51, 210–212.

Sampson, R.N., Scholes, R.J., Cerri, C., Erda, L., Hall, D.O., Handa, M., Hill, P., Howden, M., Janzen, H., Kimble, J., Lal, R., Marland, G., Minami, K., Paustian, K., Read, P., Sanchez, P.A., Scoppa, C. , Solberg, B., Trossero, M.A., Trumbore, S., Van Cleemput, O. , Whitmore, A. and Xu, D. (2000) Additional Human-induced Activities - Article 3.4. pp. 181-281 In R.T. Watson *et al.* (eds.) Land Use, Land-use Change, and Forestry. A Special Report of the Intergovernmental Panel on Climate Change. Cambridge University Press.

Schimel, D., Melillo, D, *et al.* 2000. Contributions of increasing CO_2 and climate to carbon storage by ecosystems in the United States. Science 287, 2004–2006.

Schulze, E.D., Höborg, P., Oene van, O., Persson, T., Harrison, A.F., Read, D., Kjller, A., and Matteucci, G. (2000) Interactions between the carbon and nitrogen cycle and the role of biodiversity: a synopsis of a study along a north-south transect through Europe, 468–491 in E.D. Schulze, (ed.) Carbon and nitrogen cycling in European forest ecosystems. Springer Verlag, Heidelberg.

Schulze, E.D. (Editor) (2000) Carbon and Nitrogen cycling in European forest ecosystems. Ecological Studies 142, Springer Verlag, Heidelberg.

Schulze, E.D., Wirth, C., and Heimann. M. (2000) Managing forests after Kyoto. Science 289, 2058–2059.

Tomppo, E. (1996) Multi source national forest inventory of Finland In: R. Pivinen, J. Vanclay & S. Miina (eds.), New thrusts in Forest Inventory. EFI proceedings No 7, 27–41

UN-ECE/FAO (2000) Temperate and Boreal Forest Resource Assessment 2000, Vol I, United Nations Economic Commission for Europe, Food and Agriculture Organization, Geneva, Rome, 2000, 102–142

Valentini, R., Mateucci, G, Dolman, A.J., Schulze, E.D., Rebman, C., Moors, E.J., Granier, A., Gross, P., Jensen, N.O., Pilegaard, K., Lindroth, A., Grelle, A., Bernhofer, C., Grunwald, T., Aubinet, M., Ceulemans, R., Kowalski, A.S., Vesala, T., Rannik, U., Berbigier, P., Lousteau, D., Gundmundson, J., Thorgeirsson, H., Ibrom, A., Morgenstern, K., Clement, R, Moncrieff, J., Mantagna, L. and Jarvis, P.G. (2000) Respiration as the main determinant of carbon balance in European forests. Nature 404, 861–865.

Vleeshouwers, L. M. & A. Verhagen. 2002. Carbon emission and sequestration by agricultural land use: a model study for Europe, Global Change Biology 8 (6), 519–530.

IV

IMPACT, ADAPTATION AND MITIGATION

Chapter 6

LAND-USE CHANGE, CLIMATE AND HYDROLOGY

E.J. Moors
Alterra, Wageningen University and Research Centre

A.J. Dolman
Vrije Universiteit Amsterdam

1.   Introduction

Any particular form of land use has direct effects on the way water is used in a particular area and thus also on the hydrology of that area. A particularly well-known example is the high water use of forest compared to grassland. Water use refers of course to the quantitative aspects, i.e., the availability of water. But increasingly so it also refers to the quality of the water: not all water is suited for the purpose intended. At present the standards for water quality as used in different regions, for instance in Europe, vary by several orders of magnitude. Although apparently water quality becomes more and more an issue in the political and practical sense, the main claim for water is still being made by agricultureagriculture, where at present the quantitative demand is still of more concern than quality. We therefore focus in this chapter primarily on the quantitative aspects of hydrology and land use and land-use change.

The water balance of a watershed may be written as:

$$P - E + q_{in} - q_{out} = \frac{\Delta S}{\Delta t}$$

where P is the precipitation rate, E the evaporation rate, q_{in} and q_{out} inflow and outflow rates respectively, ΔS is the change in soil water storage and Δt is the length of the period observed. Figure 6.1 shows a schematic representation of the global water cycle. Taking the total land surface of the earth as an imaginary single watershed provides the following numbers for the water balance: precipitation 111.1×10^{12} m^3 yr^{-1}, evaporation 71.4×10^{12} m^3 yr^{-1} and runoff 39.7×10^{12} m^3 yr^{-1}. The assumption is made that on a yearly basis there is no change in soil water storage.

A.J. Dolman et al. (eds.), Global Environmental Change and Land Use, 139-165.
© 2003 *Kluwer Academic Publishers.*

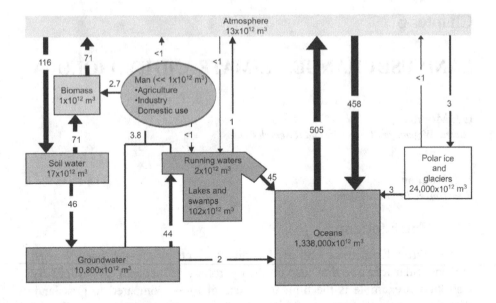

Figure 6.1. A schematic view of the global hydrological cycle. The water fluxes represented by arrows are in 10^{12} m^3 yr^{-1} (adapted from WBGU, 1999).

Mankind uses roughly 3.76×10^{12} m^3 yr^{-1} of the available fresh water. About 55% of this withdrawal is consumed, the remainder being, returned often warmed or polluted. Compared to, for example, precipitation this is a relatively small amount. However, at present the water use by man is already almost 10% of the total groundwater flow or the total surface water flow, which are the main sources of fresh water for mankind. Of the total water withdrawal agricultureagriculture accounts for about 70%, industry for 20% and domestic use for 10% (Shiklomanov, 2000). The water consumption of agricultureagriculture relative to the other economic activities is even higher: it accounts for 85% of total water use in 1995. It should be noted that the evaporation of open water bodies such as reservoirs, often constructed for irrigation or power supply, accounts for roughly 10% of total water use, i.e., more than the combined water use by industry and households. Despite the appealing simplicity of these figures, on a regional scale we find major differences in water use and withdrawals (see Figure 6.2).

The water balance of the land surface is linked strongly to a number of other balances, such as the energy, the carbon and the nutrient balance. Also, increasing the scale of the area considered for the water balance, increases the number of feedback mechanisms involved, thus complicating the analysis of the effects of simple land-use changes.

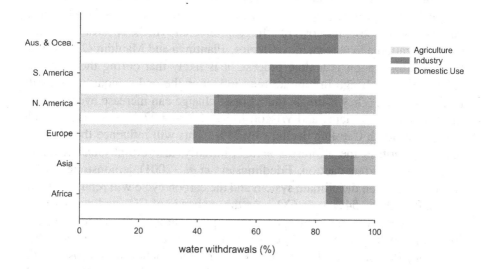

Figure 6.2. Water withdrawals as a percentage of total withdrawal (adapted from Shiklomanov, 2000).

The most important variable of the water balance is the precipitation amount. If land use or climate change reduces or increases the amount of precipitation or its distribution, it will directly effect the hydrological regime of a watershed. A number of studies have shown that a change in precipitation - if perhaps not in the annual amount, but then at least in the temporal distribution - is to be expected as a result of land use and climate change (e.g., Folland *et al.*, 2001). Determining these effects requires coupled hydrological atmospheric models, which are still in their infancy of development (see Chapter 4). Observational evidence, however, remains poor (Pitman, *et al.*, 1999).

The variable with the second most important impact on the water balance is the evaporation rate. The effects of land-use or climate change on the evaporation rate are also difficult to assess because of significant feedback mechanisms between the land-cover and evaporation rate. Again, only coupled land surface (hydrological studies) atmosphere problems can elucidate these feedback mechanisms.

Land-use change and therefore land-cover change are largely determined by human factors (see Chapters 2 and 3). For example the ratification of the Kyoto protocol allowing Annex 1 countries to use afforestation to meet CO_2 emission reduction targets may indirectly influence the climate and water balance on a regional scale. The location, as well as the area involved in afforestation, is largely determined by the costs of afforestation relative to the costs of alternative CO_2 mitigation and abatement strategies. Plantinga and Mauldin (2001) show that, as afforestation is a slow process, it is likely that during the period of afforestation a change in climate may influence the endogenous commodity prices of water. They estimate that climate change can increase average costs of afforestation in Maine and Wisconsin by 40% and 27% respectively, while the cost in South Carolina declines by 73%. This will influence the planning and realisation of the afforested area. Similarly, also bio-physical feedback mechanisms warrant caution. Friedlingstein *et al.* (2001) demonstrate that the feedback between the climate system and the carbon cycle will reduce the land uptake of CO_2 by 54% at $4 \times CO_2$. As the exchange of CO_2 and water vapour between vegetation and atmosphere takes place through the same pore, it is likely that such changes in CO_2 concentration will also effect the transpiration rate (e.g., Rabbinge *et al.*, 1993). However, other factors may constrain the carbon uptake by vegetation due to carbon dioxide fertilisation. For example Oren *et al.* (2001) showed that nutrient availability may seriously limit the positive effect on the carbon sequestration by pine forests. It is thus increasingly important to assess the effects of changes in the hydrological balance in a wider context.

The other components of the water balance, q and ΔS are largely controlled by the amount of precipitation and evaporation. Effects of land-use and climate change on these components are thus indirect. However, in- and outflow and soil water content may directly or indirectly influence land use and climate. Building dams and water diversions, canalizing riverbeds and draining wetlands are typical habitat alterations often strongly linked to land use. A reduction in soil moisture availability, which may be caused by land-use or climate change, will reduce the evaporation rate, possibly change the land-cover and thus may enhance the change in climate. Nof (2001) shows how a change in outflow indirectly caused by land-use change influences the climate of another area. He demonstrates that the construction of the Tree Gorges Dam in the Yangtze river in China and the possible associated irrigation plans, of which the latter will decrease discharge due to increased evaporation, may influence the salt content of the Japan/East Sea. This in turn may trigger deep convection in the sea, thus increasing the amount of water welling up from the deeper layers of the Japan/East Sea, changing the temperature drastically. They estimate that this temperature change will have severe impacts on the ecology of the surrounding area. Nof demonstrated similar effects but on a much smaller scale on the Mediterranean Sea due to the construction of the Aswan Dam in Egypt.

In this chapter we will focus on the effects of land-use change on the water balance and the hydrological regime of a watershed. Emphasis is also put on the influence of climate change as this itself is strongly linked to changes in land-cover. Also the feedback mechanisms of hydrology on land use and climate will be discussed. Besides the evident large impact of irrigation on the water balance, land-use change involving forestry and agricultureagriculture is in general considered to have the largest influence on the hydrological regime of a watershed. As it is also anticipated that deforestation or reforestation has an impact on climate, we will emphasise the effects of forests when considering changes in land-cover. It is not our intention to extensively review the literature associated with hydrology and land-use change and climate change; instead references are made to existing reviews. However, to elucidate aspects of uncertainty related to this subject, we will also discuss some contradictory results of studies on the effects of land-use change and climate change on hydrology.

In the next paragraphs we will first discuss the effects of land-use change and climate on the different components of the water balance of a catchment. In the final paragraph these effects are summarised and the implications for the regional and global water balance are discussed.

2. Precipitation

Precipitation is the principal factor determining water resources and water use. It is also the component of the water balance that most strongly influences the other components of the balance. The amount of precipitation depends, among others, on the land use and the associated land cover. An example of this direct link is changing agricultural land into forest. This may increase the amount of moisture in the atmosphere due to the high evaporation rate of the forest, thus enlarging the amount of precipitable water (Blyth *et al.*, 1994).

Other land use may reduce the amount of rain, for example as demonstrated by Rosenfield *et al.* (2001). They showed that clouds with a higher content of dust particles held smaller droplets, thus reducing the number of raindrops falling to the earth surface. Land use such as arable farming and grazing of cattle are the main activities that causes disruptions of the soil surface generating dust. Hence intensifying such land use may decrease the amount of rain by increasing the risk of wind erosion and thus the amount of dust in the atmosphere. These indirect effects are largely unknown.

The present general assumption is that anthropogenic climate change among others caused by land-use change will have almost no influence on the rainfall frequency. However, the extreme events are altering. This is leading towards more intense showers and longer drought periods. It should also be kept in mind that for regions with solid precipitation a temperature rise may cause a significant change in seasonal runoff. In a study on three rivers in Russia and the Ukraine reported by Shiklomanov (2000), a 2.0 to 2.5°C temperature rise caused increasing winter runoff by two to three times and decreasing spring high water runoff by 25 to 30%. Some caution should be exercised in the interpretation of these findings; for example Cubasch *et al.* (2000) show in their analysis that the uncertainty involved in the simulated change in precipitation over the next hundred years is greater than the predicted change. This uncertainty is largely due to the lack of observed data needed to evaluate the quality of simulation results.

2.1   Deforestation effects

Although deforestation has taken place over almost the whole globe, today it is primarily taking place in the tropical rainforests. A number of studies exist of the effects of deforestation of the Amazon on precipitation. In most studies a reduction of precipitation is simulated (e.g., Lean and Rowntree, 1993). Typical decreases amount to 10 to 20% of annual precipitation when the whole of the Amazon is deforested (Dolman *et al.*, 1999). This reduction is caused by a decrease in evaporation, an increase in upward long wave radiation, an increase in sensible heat flux and a reduction of the surface roughness (see also Chapter 4). However, if the deforestation is only regional, the resulting grassland may become a heat island in an otherwise moist environment and thus enhances rainfall (e.g., Avissar and Liu, 1996 and Silva Dias and Regnier, 1996). Zhang *et al.* (2001) applied a doubled CO_2 scenario combined with deforestation of the tropics to a version of the NCAR Climate Community Model. Their results showed a decrease in both precipitation and evaporation due to deforestation (see Table 6.1). Climate change modelled by doubling the CO_2 concentration gave in comparison to the decrease due to deforestation a change of a lesser magnitude in the opposite direction. Their model run with the combined land-use and climate change simulated for the Amazon Basin a change in evaporation of -179 mm yr^{-1} and in precipitation of -317 mm yr^{-1}.

In South East Asia precipitation decreased by 183 mm yr^{-1}, while over tropical Africa an increase in precipitation of 25 mm yr^{-1} was found. Thus, although most modelling studies suggest a decrease in evaporation for large-scale deforestation, at the regional scale this may be modulated to produce an increase in rainfall.

Table 6.1. Totals of the control run and changes (mm yr^{-1}) in precipitation and evaporation due to tropical deforestation, doubling CO_2 concentration and their combined effect relative to this control run (after Zhang *et al.*, 2001).

	Amazon		S.E. Asia		Africa	
	P	E	P	E	P	E
Deforestation control run	−403	−221	−241	−137	−63	−74
2×CO_2–control run	108	36	19	30	50	10
Both–control run	−317	−179	−183	−118	25	−24
Totals of the control run	1898	1243	3159	1344	1478	1023

2.2 Reforestation effects

Reforestation or afforestation may also affect climate. Harding (1992) shows that at a 10 km scale, forest will produce a small effect on the humidity, but not significant in terms of feedback to atmospheric demand downwind of the forest. However, a forest plantation at the scale of 50 km can have, in favourable synoptic situations, an influence on the generation of rainfall. A similar sensitivity of mesoscale rainfall to land-cover was demonstrated by Blyth *et al.* (1994). For an area in southwest France they showed that for a specific case of boundary conditions a full forest cover increased local rainfall by 30% compared to a bare soil. The increase in rainfall was primarily attributed to a positive feedback of evaporation of intercepted water and rainfall. This was sustained by a negative sensible heat flux above wet forest that increased the available energy for evaporation. They also hinted that increased moisture convergence in the lower levels of the atmosphere due to the enhanced surface roughness of the forest contributed to the increase in rainfall. However, care must be taken in generalising these currently available studies. To a large extent they suggest that regional differences in the forcing climate (i.e., frontal versus convective precipitation) are important in determining the final effects.

3.   Evaporation

Next to precipitation, evaporation is the other main component of the water balance of most watersheds. Evaporation may be split into three parts: - evaporation of precipitation intercepted by the canopy before hitting the soil surface (interception loss) – evaporation of water absorbed by the roots of a plant and released trough the stomata (transpiration) and – evaporation from the soil (soil evaporation). For fully vegetated surfaces the soil evaporation is in general small compared to transpiration and interception loss; for example for a mid-latitude forest the evaporation of the undergrowth is roughly 10% of total evaporation.

Sellers and Lockwood (1981) used a multi-layer crop model and a data set of one year to simulate the hydrological cycle in coniferous forest (pine), deciduous forest (oak), arable land (wheat) and grassland (see Figure 6.3).

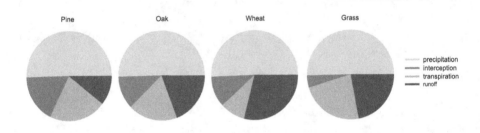

Figure 6.3.   Modelled water balance of an area in the UK with 724 mm of precipitation for four different vegetation covers (after Sellers and Lockwood,1981).

They found that on a yearly basis with a precipitation amount of 724 mm for the United Kingdom climate, the interception loss of oak and wheat are almost equal (172 and 158 mm respectively), pine losses are higher (253 mm) and those of grass lower (62 mm). Transpiration decreases from grass (335 mm) to pine (309 mm) to oak (263 mm) to wheat (134 mm). For runoff (i.e., the residual of precipitation minus evaporation) the values for oak and grass are almost equal (279 and 318 mm respectively); wheat gives the highest runoff (417 mm) and pine the lowest (152 mm). The main differences are found at the end of the growing season when the runoff for grass and wheat increases rapidly. This rapid increase is partly explained by the much deeper soil moisture depletion under forests, 22 mm of available moisture is left where grass and wheat have approximately 65 mm of moisture left at the end of the growing season.

These differences in distribution of water over the components of the water balance depend largely on the specific location. This is shown in the next example where we compare the water use of eucalypt forest and grassland. Knowledge of their water use is important in the semi-arid areas of South East Asia and Australia. The amounts of annual rainfall in Figure 6.3 and Figure 6.4 are almost identical, and so are the amounts of the other components of the water balance for the grassland area. Also the interception loss of the pine and the eucalypt trees is similar. However, the transpiration of the eucalypt trees is about eight times as high as that of the pine. The low evaporation rate of grass led to rising of saline water tables after changing forest to agricultural land in Australia (Greenwood, 1992). After a number of years this rendered the land useless for agricultureagriculture.

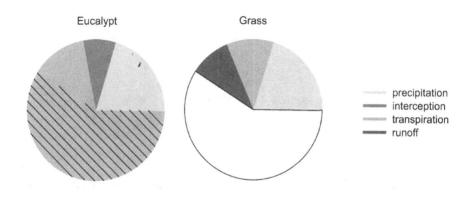

Figure 6.4. Measured water balance for eucalypt trees and grazed grassland in Australia with 684 mm of precipitation (after Greenwood, 1992). The hatched area shows the amount of water extracted from the groundwater store to maintain transpiration.

There are also interannual differences in the ratio of the different components of the water balance. Figure 6.5 shows an example for a pine forest in the Netherlands. At this specific site the differences between the years are not only caused by the differences in total precipitation amount, but also by the seasonal distribution of the precipitation and the radiation load as well as the length of the growing season. These last two factors are the main variables influencing the yearly evaporation amount at this site.

Forest evaporation may also change between years due to ageing of the trees. For example Murakami *et al.* (2000) show that forest evaporation (that is transpiration plus interception plus evaporation of the undergrowth and soil) increases with age and remains stable after canopy closure. They used rainfall and runoff data to estimate evaporation. They compared data of a mature stand from 1981 to 1985 and a young stand from 1991 to 1994. For their stands they modelled the evaporation peak at 20 years. They claim that transpiration is the main source of variation in forest evaporation with age. Although some care should be taken as their conclusions on interception are based on only two years of observations in a mature forest. They summarise data from mainly Japan and the Soviet Union on forest age associated with transpiration. The lowest transpiration was found for ash and aspen stands aged 20–40 years and the highest for oak and spruce stands aged 40–60 years. Murakami *et al.* (2000) also mention that Yuruki (1964) found that even a young forest of four years old can reach its maximum leaf biomass if the canopy completely closes. In the Netherlands with its in general intensive forest practice, the management is such that for slow and fast growing trees a planting density is chosen to ensure canopy closure in just a few years (two to six years).

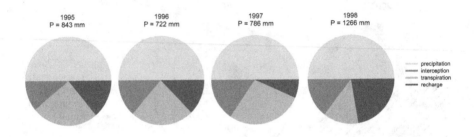

Figure 6.5.   Measured water balance of a pine forest in the Netherlands for four different years.

3.1   Potential evapotranspiration

For modelling purposes, often the concept of potential evapotranspiration is used. This is somewhat incorrect, as total evaporation consists of both transpiration (through the stomata) and evaporation from soil and wet canopy. In practice these are often lumped together under the name evapotranspiration, thereby losing the distinction, and possibility to distinguish, between two radically different processes. McKenney and Rosenberg (1993) predicted an increase in potential evapotranspiration by about 2 to 3% per °K temperature rise provided that there is no change in down-dwelling solar radiation and in the surface resistance. Martin (1989) reported an increase in evaporation of 2% per °K, for grassland and of 8% per °K for forest. These values are only valid for stable environmental conditions.

A typical deforestation scenario in climate change modelling is represented by a reduction in minimum leaf area index (from 5.0 to 0.5), a reduction in surface roughness (from 2.0 m to 0.2 m), an increase in albedo (from 12% to 19%), a brighter soil colour and a coarser texture (Zhang *et al.*, 2001). At all three tropical regions modelled Zhang *et al.* (2001) found an increase in temperature of more than 2.0 K. This large increase in surface temperature was not only thought to be due to the doubled CO_2 concentration, but also because of the reduction of evaporation (caused by deforestation) and the increase of downward atmospheric long wave radiation. Using typical increases in evaporation per °K this would lead to an increase of evaporation loss of more than 16% for forest. However, in general the stomatal resistance will limit the transpiration rate. The degree of opening of the stomata is considered as a compromise in the balance between the limitation of water loss and admission of CO_2. Rabbinge *et al.* (1993) estimate that the doubling of CO_2 may suppress transpiration from C3 plants (such as trees and temperate grasses) by 10–20% and from C4 plants (warm-zone forage grasses) by 25%. Mooney *et al.* (1999) show that there is a wide range varying from negative to +85% in the increase of above ground biomass for grasslands if exposed to double ambient CO_2. This variation in response to the elevated CO_2 level shows the highly interactive nature of the CO_2 response with other environmental factors such as water and nutrient availability and temperature. In accordance with the findings of Rabbinge *et al.* (1993), Mooney *et al.* (1999) conclude from experimental data sets that herbaceous plants as well as tree seedlings exposed to elevated CO_2 show a reduction in stomatal conductance. However, this was not the case for mature trees (forests).

Due to readily available temperature data and the difficulty to obtain radiation, wind speed and humidity data, studies on the effects of climate change are often based on potential evaporation equations using temperature as the driving variable. Often these equations do not take into account feedback mechanisms between these variables and the physiology of the plant (for example De Wit *et al.*, 2001). Such equations may seriously overestimate the evaporation loss under different climatic conditions. Thus care should be taken in using potential evaporation equations based only on the temperature as the driving variable and not taking into account the possibility of a change in stomatal conductance.

Lockwood (1999) suggests that the suppression of potential evapotranspiration by enhanced CO_2 levels will be small, but that actual transpiration from tall, slow-growing vegetation covers may be significantly suppressed. This suppression will be most noticeable in dry climates, and will be masked in very wet climates where a large proportion of evaporation consists of interception loss. Due to the complicated feedback mechanisms a clear indication of the effects on interception loss is difficult.

3.2   Interception loss

Murakami *et al.* (2000) claim that for mature forests (i.e., canopy closure reached) in contrast to young forests, interception loss increases with the amount of rain. The relation between interception loss and the amount of rain is to be expected as at the locations of their study there is a strong relation between the amount of rainfall and the duration of rain. The higher interception loss of the mature forest relative to the younger forest is explained by a higher aerodynamic resistance for younger forest. Jarvis *et al.* (1976) showed that the minimum aerodynamic resistance is already reached at a tree height of 4 m. It should be noted that some tree species may reach this height already after two years, thus minimising differences in interception losses due to stand age. The tree spacing also determines the magnitude of the aerodynamic resistance. Teklehaimanot *et al.* (1991) found that the annual interception loss as a percentage of gross rainfall was 33, 24, 15 and 9% in Sitka spruce stand with a tree spacing of 2, 4, 6 and 8 m respectively.

Changes in the precipitation regime will most likely be accompanied by a change in other climate variables. However, ignoring possible large scale feedback mechanisms and assuming that increased CO_2 levels will have a negligible effect on leaf area of natural vegetation, it is possible to give some general indications of the change in interception loss due to a change in precipitation. Increased rainfall intensity will have a reducing effect on interception loss expressed as a percentage of gross rainfall. Also a reduction of the number of showers in a year will also reduce interception loss. Figure 6.6a. shows the interception loss per month of a pine forest in the Netherlands as a function of the number of days binned for different daily precipitation rates. Rainfall events with an intensity between 10 and 20 mm d^{-1} generate the highest interception loss on a monthly basis. However, the interception loss as a percentage of rainfall is highest for precipitation rates of less than 5 mm d^{-1}. At this location, days with such low rainfall intensities are much more frequent than higher precipitation rates. A shift in the frequency distribution of rainfall to higher intensities may therefore increase the total amount of interception loss, but will most likely reduce the interception loss as a percentage of rainfall.

4. Runoff

Running water is the main source of fresh water. The flow of a river may be divided into two parts: – peak flow, which is a relatively flashy and short phenomenon and – base flow, which is smaller in volume but has a more regular flow throughout the year. On a relatively long time scale, runoff is the result of the difference between rainfall and evaporation. In some areas with low precipitation amounts runoff may be non-existent, while in others, often during short periods, the size of the flow is such that whole areas are flooded.

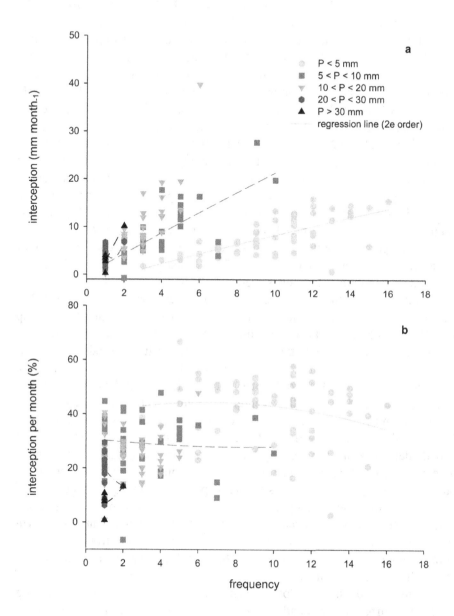

Figure 6.6. Interception loss per month as a function of the number of days at a pine forest site in the Netherlands binned for different rainfall intensities (graph a). Bottom graph is the same as the top graph but with the interception loss as a percentage of the precipitation amount (graph b).

4.1 Floods

In the existing hydrological data records it is difficult to detect any signifi-cant trend in, e.g., rainfall and flooding that can be attributed solely to long-term changes in climate or land use (Robson *et al.*, 1998). Matheussen *et al.* (2000)

studied the effects of the change of the historical land-cover from 1900 to the present 1990 land-cover on streamflow for the interior Columbia River Basin in the USA and Canada. Their results show that the most important change in hydrology related to land-use change has been a general tendency towards decreased vegetation maturity in the forested areas of the basin. This trend represents a balance between logging and fire suppression. Using the Variable Infiltration Capacity (VIC) model they found an increase in annual average runoff varying between 3.1 and 12.1% and a decrease in evapotranspiration ranging from 3.1 to 12.1% relative to the 1900 vegetation cover for the different sub-basins. Reynard *et al.* (2001) used a continuous flow simulation model (CLASSIC) to assess the potential impact of climate and land-use change on the flood regimes of large watersheds in the UK. Their climate change scenarios are based on the HadCM2 experiments from the Hadley Centre. These experiments use a 1% increase in CO_2 concentration a year after 1990 and suggest a global temperature increase of about 3K in 2100. For the 2050s the mean of these scenarios changed rainfall to +20% in January and −19% in July. For potential evaporation the changes were −25% in January and +20% in July. The combination of these changes in rainfall and evaporation generally led to wetter winters and drier summers. This resulted in increased winter flows by up to 20% with the highest flow occurring more frequently and a bigger increase of the extreme flood events than those with lower return periods. The rainfall intensity and distribution mainly determine the hydrological response of a watershed. As there is less confidence in the daily rainfall data of the GCMs, Reynard *et al.* (2001) used the aggregated monthly data. The method of applying a monthly rainfall scenario to a daily hydrological model is important. Therefore three methods were used to convert monthly rainfall from the GCM to daily data. Keeping the monthly total the same for all methods, the methods reflect: a proportional increase for each rain day in the month, an increase in rainy and dry days in winter and summer respectively and an 'enhanced storm' profile. The increase in rainy days caused only a small increase in the flood frequency curve under all scenarios. The effect of the enhancement of the storm events as applied in their study was almost indistinguishable from the proportional change in rainfall. This was thought to be due to a low threshold (5 mm) chosen for the enhancement.

Also three land-use scenarios were applied: a best guess scenario increasing the existing urban areas (5 and 6% coverage) by 2%, an increase to an overall urbanisation of 15%, a reforestation scenario to produce a 50% forest cover. The best guess land-use scenario did not produce significant differences, however the effect of the major increase in forest cover is large enough to fully compensate for the shifts due to climate in some scenarios. Increasing the urban cover increases the winter flows and enhances the effect due to climate.

4.2   Low flows

The occurrence and the duration of low flows in general determine the lower limit for the amount of water available for human use. Low flows are not only representative for the state of the surface water, but as the base flow is usually the result of subsurface and or groundwater flow, it is also an indication of the state of the groundwater reservoir. Another problem often strongly related to low flows is the quality of water. Due to the small volume of water under such conditions, the contaminant load is high and may become hazardous if it cannot be controlled during periods of extreme low flows.

De Wit *et al.* (2001) show in a study on the Meuse watershed comprising France, Belgium and the Netherlands that the possible decrease in summer discharge (low flow) is not in accordance with the apparent increase of the summer precipitation in the Ardennes. They suggest that the decrease might be the result of changes in human water withdrawals, flow regulation and/or changes in land use. However, due to the high natural variability of the discharge of the Meuse and the limited length of the record, no definite conclusions could be drawn. To investigate the possible effect of future changes in the watershed, some case studies were done using different models to simulate the behaviour of some sub-watersheds.

For a relatively flat sub-watershed in the Netherlands, a total stop of irrigation by sprinkling would decrease evaporation and discharge in summer time by 5-10% if compared to the situation today. For the same watershed, removal of all field drains (at present 5% of the area is drained) and field ditches (on average there is a ditch every 180 m in the study watershed) decreases the winter discharge and increases the summer discharge. The overall effect (see Figure 6.7) of these two changes is an increased evaporation of more than 20% in the summer months, a reduction of the high winter flows as well as an increase in the low summer flow (i.e., groundwater flow). These changes in water balance terms compare in magnitude to the expected changes resulting from climate induce changes.

For an upstream watershed with 7% of the area forested, complete afforestation using deciduous trees to 100% coverage increases evaporation especially in April-May (>10%) and August-September (>20%). The monthly average discharge reduced during the year by more than 5% in March to 20% in October (see Figure 6.8).

From these simulations it may be expected that, at least for the temperate zone, changes in flow regulation and land use (possibly indirectly the result of climate change) have more impact than changes in precipitation and evapora-

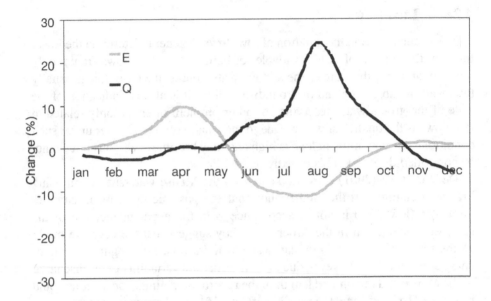

Figure 6.7. Impact of a stop in irrigation and drainage on the actual evapotranspiration (E) and discharges (Q) in the Beerze watershed. Change represents the relative difference (%) between the baseline run and the scenario run.

tion solely due to direct climate changes. However, this may not be the case in regions prone to drought such as the Mediterranean and Sahelian region, where even relatively small changes in precipitation may have drastic effects on land use. For all regions the water quality, which is severely under pressure during extremely dry and warm periods, will be more influenced by changes in emissions and changes in flow regulation than by changes in climate.

4.3 Change in water yield due to land-cover change

Brattsev (1979) pointed out that it is difficult to assess the effects of land-cover change using a single watershed. The uncertainty associated with such research may be demonstrated using the study of Putuhena and Cordery (2000). They studied the effect of changing the vegetation cover from eucalypt to Pinus radiata. Their experimental watershed is situated in NSW Australia, has an area of 9.4 ha and an average slope of 12%. They measured stream flow and precipitation continuously. Throughfall and stem flow was measured for two years and 13 months respectively, starting in 1993. They estimate the average annual interception loss of the canopy for a mature eucalypt forest at 14.9% of gross rainfall, while the interception loss of the forest floor is estimated at 7% of gross rainfall. During a growing period of one year old to a 16 year old pine forest, the interception loss was estimated to change from 1.1 to 27.0%

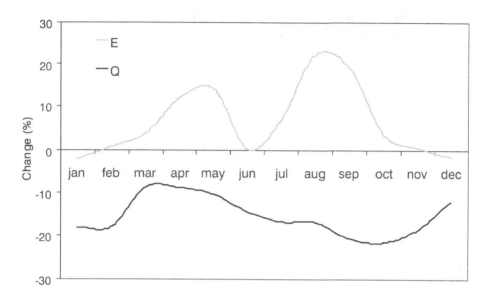

Figure 6.8. Impact of a complete afforrestation (decidious forest) on the actual evapotranspiration (E) and discharges (Q) in the Mehaigne watershed. Change represents the relative difference (%) between the baseline run and the scenario run.

for the canopy and from 0.4 to 11.7% for the forest floor. From their primarily modelling results they conclude that there is a clear reduction of the stream flow during this 16 year growing period. This reduction is mainly caused by the increasing percentage of interception losses. They also show a slight increase of the sum of transpiration and soil evaporation from 67 to 75% of gross rainfall during the first four years after changing the vegetation type. They suggest that the expected decrease in tree transpiration is offset by an increase in soil evaporation and transpiration of opportunistic ground cover vegetation due to an increased radiation load.

The top graph of Figure 6.9 shows the runoff coefficient (runoff as a fraction of precipitation) for each year. Looking at the period after 1978 the runoff coefficient decreases. However, looking at the lower graph where the runoff coefficient is plotted against the yearly precipitation, one can see that there is no clear evidence for a decrease of the runoff coefficient after changing the vegetation cover. Arguably, this is due to the first number of years just after the change from eucalypt to Pinus radiata when the runoff coefficient is relatively high. Also the increase of the runoff coefficient during the years with eucalypt cover is in contrast to the conclusion that runoff decreases with tree growth, as could be concluded from the Pinus radiata data. This shows the difficulty in interpreting the results of watershed studies on the effects of land use or vegetation change on the hydrological regime.

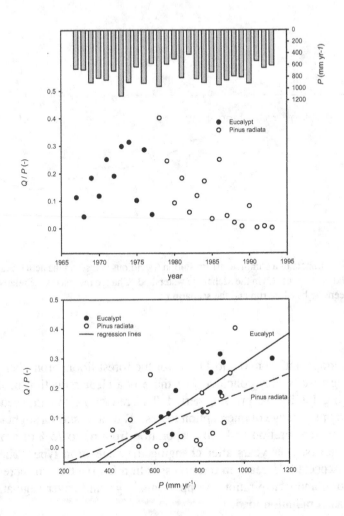

Figure 6.9. Effects of harvest of eucalypt trees followed by planting of Pinus radiata on the runoff coefficient (top graph). The lower graph shows the scatter plot of the runoff coefficient as a function of precipitation.

A better result is obtained using paired watershed experiments. An example of this is reported by Hibbert (1967). He compared two small forested basins in the USA. One watershed with natural vegetation consisting mainly of oak-hickory was clearcut. After a sharp increase of streamflow in the first year, it decreased logarithmically until full cover was reached in approximately 23 years. A second clearcutting at that time gave the same response.

Although most studies are on deforestation or reforestation, also afforestation seems to give similar responses. Lill *et al.* (1980) showed by simple regression a reduction in streamflow of 300 to 380 mm yr^{-1}, from the third year after afforesting a natural grassland by planting Eucalyptis grandis in South Africa.

Bosch and Hewlett (1982) reviewed 94 watershed experiments to determine the effect of vegetation change on water yield. They concluded that pine and eucalypt forest types cause on average 40 mm change in water yield per 10% change in cover, deciduous hardwood 25 mm and scrub 10 mm. It should be noted that most of these watersheds were located at higher altitudes and are mostly relatively small head water watersheds.

4.4   Change in peakflows due to land-cover change

Beschta *et al.* (2000) give an overview of studies for the western Cascade Range of Oregon, USA done on the effects of clearfelling and shelterwood silviculture on the peakflows using paired small-watersheds. They distinguished between small peakflows occurring a couple of times a year and large peakflows with an occurrence interval of more than a year. The size of the basins varied between 10 and 100 ha and the percentage of the area harvested varied between 25 and 100%. For the small peakflows for three watersheds an increase was reported while for four watersheds no change was reported after clearfelling. On the single watershed where harvesting by shelterwood silviculture was done showed no change afterwards. For large peakflows, studies on four watersheds reported no change after harvesting, while studies on three watersheds showed an increase. An interesting detail is the fact that two studies on the same watershed showed opposite results. This may be caused by the observation that the study showing an increase used a recurrence interval greater than 0.4 years instead of more than one year, as was done in the other studies. The study on one watershed reported an increase while another reported no change for large peakflows after harvesting by shelterwood silviculture. After reanalysis of the data of some of the watersheds, Beschta *et al.* (2000) concluded that in the small basins (<101 ha) peakflows with a return period varying between 0.4 and 5 yr increased. However, this is not evident for peakflows with a return period of 5 yr or more. For large basins (>60 km^2) they did not find strong evidence for increasing peakflows after harvesting.

5.   Soil water

The characteristics and the state of the soil largely determine the evaporation rate and the partitioning of the precipitation into direct runoff and water infiltrating into the soil. A well-known effect of land use influencing the characteristics of the top soil is the reduction of the infiltration rate due to soil crusting (scaling) caused by for example deforestation and erosion as encountered in some areas of the Sahelian countries. Besides the change in the infiltration rate of the topsoil, deforestation may also lead to the loss of macropores in fine textured soils. Thus changing land use from forest to e.g., grassland or vice versa may well change the hydrological behaviour of a watershed. An important parameter determining how these changes work out is the hydraulic conductivity of the soil. In general a higher hydraulic conductivity of the topsoil is found under forest if compared to grassland, although this is not the case for all studies. In the Netherlands, forest growth on the newly claimed land in the polders caused in lower parts of the clay soils the formation of permanent cracks with a width of up to 10 cm. Under grassland, cracks of this extent were not found. The cause of this is the combination of a deeper rooting depth and a higher evaporation rate of forest compared to agricultural land use These cracks function as natural drains and partly counterbalance the reduction in peak flow caused by the increased evaporation rate.

Changes in infiltration rate caused by land-use change may well explain the sometimes reported conflicting effects of land-use change on water yield. For example if deforestation does not effect the infiltration rate, the reduced evaporation rate may well cause an increase in base flow. However, if forest removal reduces the infiltration rate this may well increase the peakflow. Thus the increased peakflow will leave less water for the base flow even with a decreased evaporation rate (Bonell, 1998).

The effect of soil water on the global cycle is demonstrated by Dolman *et al.* (2001). They replaced in a 3-D atmospheric model a non-variable soil map and a Clapp and Hornberger parameterisation by a variable soil map and a Mualem–van Genuchten parameterisation. This caused shifts in the modelled evaporation, precipitation and runoff. A qualitative description of these changes is found in Table 6.2. The relatively low hydraulic conductivity of the European soils as a whole generates in comparison to the non-variable soil map more precipitation, increases evaporation and surface runoff and reduces base flow.

6.　　Implications for the regional and global scale

In the preceding paragraphs the effects of land-use change and climate on the different components of the water balance of a catchment were discussed. Here these effects are summarised and the implications for the regional and global water balance are discussed.

Table 6.2. The changes resulting from the implementation of the variable soil map, based on Mualem–van Genuchten parameterisation, in comparison to the non-variable soil map based on Clapp and Hornberger parameterisation (Dolman *et al.*, 2001). RH = relative humidity and T = temperature.

	Area	Climate type	Soil water balance	General conclusions
Europe (total)		–	More precipitation and evapotranspiration in summer. More surface runoff and less deep runoff.	Small hydraulic conductivity (k) and large soil water storage leads to more continental evapotranspiration and precipitation in summer. RH and T confirm this.
Coarse	Norway	Oceanic/Sub-arctic (wet)	More deep runoff and less evapotranspiration in summer.	In wet conditions, k is larger than the Clapp and Hornberger soil. This leads to less evapotranspiration and more runoff. Model error for RH and T doesn't decrease much however.
	Poland	Continental (dry)	More precipitation in summer, and little more evapotranspiration. More deep runoff in wet conditions.	
Medium	Sweden	Sub-arctic (wet)	No important changes in precipitation and evapotranspiration. More surface runoff in spring, less water storage in soil and thus less deep runoff in summer.	More soil water storage capacity and smaller k results in intensification of the mainland circulation of evapotranspiration and (particular convective) precipitation. Annual amplitude of soil water content in the upper soil layers with highest plant root density has increased. Deep runoff decreases, but frequency of surface runoff increases. Enhanced RH and lowered T for area 5 support the increase of evapotranspiration.
	Spain	Mediterranean (dry)	Less deep runoff in winter and significantly more evapotranspiration in summer.	
Medium-Fine	France Germany	Altered Oceanic	More convective precipitation and evapotranspiration in summer	More deep runoff at field capacity, as hydraulic conductivity is relative large.
Fine + Very Fine	Hungary	Continental	Slightly increased summer evapotranspiration and surface runoff	Too little data for reliable analysis. Lower hydraulic conductivity and increased storage cause probably more surface runoff and evapotranspiration.
Histosols	UK Ireland	Oceanic	Evidence for more evapotranspiration and less deep runoff in summer.	

Future land use and the associated water withdrawals and water use are strongly associated with population growth. Predictions made in 1980 showed that care should be exercised interpreting these predictions. These earlier predictions were almost completely based on population predictions, and showed a steep increase in water withdrawals for the 1990s and the year 2000 that was never realised. Although at present more feedback mechanisms as well as expected technical developments are incorporated in models used to predict future water use and water withdrawals, the uncertainty in the results of these models remains high.

Figure 6.10 shows the effect of a 'best guess' scenario on the water withdrawals in 2025 presented in WBGU (1999). This compares relatively well with the estimate made by Shiklomanov (2000), except for the predicted industrial withdrawals. In the report of WBGU (1999) it is assumed that private households have a continuous increase in water use for annual incomes up to US$15,000. Followed by a rapid decline to 50% of the peak value, where it remains also at higher incomes. Industrial withdrawals depend on the availability of water. These withdrawals remain at a constantly high level for incomes from US$5,000 to US$15,000. This is followed in both cases by a rapid decline to 50% of the initial level. It is also assumed that the irrigation efficiency increases, and that in most developing countries an increase in population is followed by additional areas being brought under irrigation.

For comparison also the projection for 2025, solely based on population, is added. The main differences found for agricultural water use are due to the stagnation of the population growth and an improvement of the irrigation efficiency. This will reduce the water use in the more industrialised countries, whereas in most countries of Africa and South America the increase in withdrawals for agricultureagriculture will continue at a high rate. The improvement of the irrigation efficiency, now often less than 40% for gravity irrigation systems may well cause a new problem in some areas due to salinisation of the top soil. If this is difficult to reverse, it may well put an extra demand on the available land for agricultureagriculture.

From these three categories, agricultureagriculture combines major demands on both water and land use. A potential 110 million hectares was estimated in 1990 as still available for irrigation (World Bank and UNDP, 1990). If population increases at the same rate as between 1980 and 1990 this will totally be in use by 2025 (World Bank and UNDP, 1990). Even taking on board the relatively high uncertainty of these predictions, it is clear that for the coming years the demands on the available water resources are only increasing. At the same time, the number of stakeholders involved is increasing, complicating the

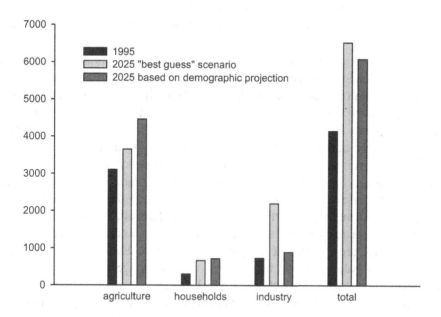

Figure 6.10. The water withdrawals as predicted for 2025 using the 'best guess' scenario and based on the demographic projection. As a reference the withdrawals in 1995 are depicted (based on WBGU, 1999).

distribution of water among them. It is therefore crucial to accurately plan and manage our water resources. This requires insight in all processes involved, both the physical, as mainly described in this chapter, as well as the socio-economic factors shaping future demand.

As agricultureagriculture is the main water consumer, and as land-use changing from agricultureagriculture to forestry and vice versa has the biggest effect on hydrology, some more specific conclusions may be drawn. Firstly, the impact on hydrology due to rigorous land-use changes can be much larger (on average four times as large (Zhang *et al.*, 2001)) as the impact due to climate change. Secondly, changes in precipitation due to land-use change may be expected if relatively large areas ($>50 \times 50$ km^2) are involved. This is especially the case if the land-use change is accompanied by a change in vegetation cover due to re- or deforestation. In general forests have a positive feedback to downwind precipitation rates. The two main causes of these differences are a higher evaporation rate of intercepted rainfall and a deeper rooting depth of forest. However, favourable synoptic conditions are necessary to enable this process.

With respect to the water yield of watersheds, the general effects due to land-cover change may be summarised as follows. Deforestation may result in considerable increase in water yield, and the highest increase is expected to take place in the first year after the change in vegetation cover. The size of the change depends on the replacement vegetation cover. This increase in water yield is mainly caused by delayed flow, and less by storm runoff. Changes in vegetation cover will have less or no influence on peakflows with long return periods (e.g., >1 yr). These general effects due to land-use change may be strongly influenced (even contradicted) by changes in the surface infiltration capacity also resulting from the land-use change (Bonell, 1998).

In general, natural ecosystems are well balanced and all components of the water balance are in a sort of long-term equilibrium. Relatively small effects on the water balance are to be expected due to changes in evaporation and precipitation possibly indirectly caused by land-use change. However, the main effects on the water balance are due to human activity, not only agriculture, but also hydrological measures that are often involved to facilitate a certain land-use.

References

Avissar, R. and Liu, Y. (1996) Three-dimensional numerical study of shallow convective clouds and precipitation induced by land surface forcing. J. Geophys. Res. 101 (D3), 7499–7518.

Beschta, R.L., Pyles, M.R., Skaugset, A.E. and Surfleet, C.G. (2000) Peakflow responses to forest practises in the western cascades of Oregon, USA. J. Hydrol. 233, 102–130.

Blyth, E.M., Dolman, A.J. and Noilhan, J. (1994) The effect of forest on mesoscale rainfall. An example from HAPEX-MOBILHY. J. Appl. Meteorol. 33, 445–454.

Bonell, M. (1998) Possible impacts of climate variability and change on tropical forest hydrology. Climate Change 39, 215–272.

Bosch, J.M. and Hewlett, J.D. (1982) A review of catchment experiments to determine the effect of vegetation changes on water yield and evapotranspiration. J. Hydrol. 55, 3–23.

Brattsev, S.A. (1979) Hydrologic role of the forest in the Komi ASSR. Soviet Hydrology: Selected Papers. Vol. 18 No 2, 127–133.

Cubasch, U., Voss, R. and Mikolajewiscz, U. (2000) Precipitation: A parameter changing climate and modified by climate change. Climatic Change 46, 257–276.

De Wit, M., Warmerdam, P., Torfs, P., Uijlenhoet, R., Roulin, E., Cheymol, A., van Deursen, W., van Walsum, P., Kwadijk, J., Ververs, M. and Buitenveld, H. (2001) Effect of climate change on the hydrology of the river Meuse. NOP-report (in prep.)

Dolman, A.J., Silva Dias, M. A., Calvet, J.C., Ashby, M., Tahara, A.S., Delire, C., Kabat, P., Fisch, G.A., Nobre, C.A. (1999) Meso-scale effects of tropical deforestation in Amazonia: preparatory LBA modelling studies. Ann. Geophysicae 17, 1095–1110.

Dolman, A.J., Soet, M., Van den Hurk, B.J.J.M., Ijpelaar, R.J.M. and Ronda, R.J. (2001) The representation of the seasonal hydrological cycle in a regional climate model in West Europe. In: Dolman, A.J., Hall, A.J., Kavvas, M.L., Oki, T. and Pomeroy, J.W. (eds.). Soil-Vegetation-Atmosphere Transfer Schemes and Large-Scale Hydrlogical Models. IAHS Publication No. 270, 11–18.

Folland, C.K., Karl, T.R., Christy, J.R., Clarke, R.A., Gruza, G.V., Jouzel, J., Mann, M.E., Oerlemans, J., Salinger, M.J. and Wang, S.-W. (2001) Observed Climate Variability and Change. In: Houghton, J.T., Ding, Y., Griggs, D.J., Noguer, M., van der Linden, P.J., Dai, X., Maskell, K. and Johnson, C.A. (eds.). Climate Change 2001: The Scientific Basis. Contribution of Working Group I to the Third Assessment Report of the Intergovernmental Panel on Climate Change. Cambridge University Press, 99–181.

Friedlingstein, P., Bopp, L., Ciais, P., Dufresne, J.-L., Fairhead, L., LeTreut, H., Monfray, P. and Orr, J. (2001) Positive feedback between future climate change and the carbon cycle. Geophysical Research Letters 28 (8), 1543–1546.

Greenwood, E.A.N. (1992) Water use by eucalypts-measurements and implications for Australia and India. In: Calder, I. R., Hall, R. L. and Adlard, P. G. (eds.). Growth and Water Use of Forest Plantations, Wiley, Chichester, UK, 298–300.

Harding, R.J. (1992) The modification of climate by forests. In: Calder, I.R., Hall, R.L. and Adlard, P.G. (eds.). Growth and Water Use of Forest Plantations, Wiley, Chichester, UK, 332–346.

Hibbert, A.R. (1967) Forest treatment effects on water yield. In: Sopper, W.E. and Lull, H.W. (eds.). Forest Hydrology. Pergamon Press, Oxford, 536–538.

Jarvis, P.G., James, G.B., and Landsberg, J.J. (1976). Coniferous Forest. In: Monteith, J.L. (ed.), Vegetation and the Atmosphere 2. Academic Press, London, 171–240.

Lean, J. and Rowntree, P. (1993) A GCM simulation of the impact of Amazonian deforestation on climate using an improved canopy representation. Q.J.R. Meterol. Soc. 119, 509–530.

Lill, V.W.S., Kruger, F.J. and Wyk, V.D.B. (1980) The effect of afforestation with Eucalyptus Grandis Hill Ex Maiden and Pinus Patula Schlecht-Et Cham. On streamflow from experiment catchments of Mokobulaan, Transvaal. J. Hydrol. 48, 107–118.

Lockwood, J.G. (1999) Is potential evapotranspiration and its relationship with actual evapotranspiration sensitive to elevated atmospheric CO_2 levels? Climate Change 41, 193–212.

Martin, P. (1989) The significance of radiative coupling between vegetation and the atmosphere. Agric. For. Meteorol. 49, 45–53.

Matheussen B., Kirschbaum, R.L., Goodman, I.A., O'Donnell, G.M. and Lettenmaier, D.P. (2000) Effects of land cover change on streamflow in the interior Columbia River Basin (USA and Canada). Hydrol. Process. 14, 867–885.

McKenney, M.S. and Rosenberg, N.J. (1993) Sensitivity of some potential evapotranspiration estimation methods to climate change. Agric. For. Meteorol. 64, 81–110.

Mooney, H.A., Canadell, J., Chapin III, F.S., Ehleringer, J.R., Korner, Ch., McMurtrie, R.E., Parton, W.J., Pitelka, L.F. and Schulze, E.D. (1999) Ecosystem physiology responses to global change. In: Walker, B., Steffen, W., Canadell, J. and Ingram, J. (eds.). The terrestial biosphere and global change. Implications for natural and managed ecosystems. International geosphere-biosphere programme book series. Cambridge University Press.

Murakami, S., Tsuboyama, Y., Shimizu, T., Fujieda, M. and Noguchi, S. (2000) Variation of evapotranspiration with stand age and climate in a small Japanese forested catchment. J. Hydrol. 227, 114–127.

Nof, D. (2001) China's development could lead to bottom water formation in the Japan/East Sea. Bull. American Meteorological Society 82, 609–618.

Oren, R., Ellsworth, D.S., Johnsen, K.H., Phillips. N., Ewers, B.E., Maier, C., Schfer, K.V.R., McCarthy, H., Hendrey, G., McNulty, S.G. and Katul. G.G. (2001) Soil fertility limits carbon sequestration by forest ecosystems in a CO_2-enriched atmosphere. Nature 411, 469–472.

Pitman, A., Pielke, R., Avissar, R., Claussen, M., Gash, J. and Dolman, A.J. (1999) The role of land surface in weather and climate: does the land surface matter? IGBP-Global Change Newsletter 39, 4–11.

Plantinga, A.J. and Mauldin, T. (2001) A method for estimating the cost of CO_2 mitigation through afforestation. Climate Change 49, 21–40.

Putuhena, W.M. and Cordery, I. (2000). Some hydrological effects of changing forest cover from eucalypts to Pinus radiata. Agric. For. Meteorol. 100, 59–72.

Rabbinge, R., van Latesteijn, H.C. and Goudriaan, J. (1993) Assessing the Greenhouse Effect in Agriculture. In: Lake, J.V., Bock, G.R., and Ackrill, K. (eds.). Environmental Change and Human Health, Ciba Foundation Symposium 175, John Wiley, Chichester, 62–79.

Reynard, N.S., Prudhomme, C. and Crooks, S.M. (2001) The flood characteristics of large U.K. rivers: potential effects of changing climate and land use. Climatic Change 48, 343–359.

Robson, A.J., Jones, T.K., Reed, D.W., and Bayliss, A.C. (1998) A study of national trend and variation in U.K. floods. Int. J. Climatol. 18, 165–182.

Rosenfield, D., Rudich, Y. and L. Ronen (2001) Desert dust suppressing precipitation: A possible desertification feedback loop. Proceedings of the National Academy of Science 98, 5975–5980.

Sellers, P.J. and Lockwood, J.G. (1981) A numerical simulation of the effects of changing vegetation type on surface hydroclimatology. Climate Change 3, 121–136.

Shiklomanov, I.A. (2000) World Water Resources: Modern Assessment and Outlook for the 21st Century. IHP/UNESCO, Paris, France.

Silva Dias, M.A.F. and Regnier, P. (1996) Simulation of mesoscale simulations in a deforested area of Rondonia in the dry season. In: Gash, J.H.C., Nobre, C.A., Roberts, J.M. and Victoria, R.L. (eds). Amazonian deforestation and climate. Wiley & Sons, 531–548.

Teklehaimanot, Z., Jarvis, P.G. and Ledger, D.C. (1991) Rainfall interception and boundary layer conductance in relation to tree spacing. J. Hydrol. 123, 261–278.

WBGU (German Advisory Council on Global Change) (1999) World in transition: ways towards sustainable management of freshwater resources. Springer-Verlag Berlin-Heidelberg.

Yuruki, T. (1964) Analytical studies on factors controlling tree growths. Bull. Kyushu Univ. For. (Fukuoka, Japan) 37, 85-178 (in Japanese with English summary).

Zhang, H., Henderson-Sellers A. and McGuffie K. (2001) The compounding effects of tropical deforestation and greenhouse warming on climate. Climate Change 49, 309–338.

Chapter 7

CLIMATE CHANGE AND FOOD SECURITY IN THE DRYLANDS OF WEST AFRICA

A. Verhagen
Plant Research International, Wageningen University and Research Centre

A.J. Dietz
Amsterdam Research Institute for Global Issues and Development Studies, University of Amsterdam

R. Ruben
Development Economics Group, Department of Social Sciences, Wageningen University and Research Centre

H. van Dijk
African Studies Centre Leiden

A. de Jong
Institute of Development Studies, Faculty of Geographical Sciences, University of Utrecht

F. Zaal
Amsterdam Research Institute for Global Issues and Development Studies, University of Amsterdam

M. de Bruijn
African Studies Centre Leiden

H. van Keulen
Plant Research International, Wageningen University and Research Centre

A.J. Dolman et al. (eds.), Global Environmental Change and Land Use, 167-185.
© 2003 *Kluwer Academic Publishers.*

1.   Introduction

Increasing population has prompted farmers to claim land for agriculturea-griculture and use this land more intensively. This strategy was successful in feeding the world's population but came with a cost. These environmental costs became clear over the last few decades: degradation of once fertile land, pollution of water resources, expansion of agriculture into marginal areas, conversion of natural systems to agricultural systems resulting in loss of biodiversity. Currently, approximately 38% of the total land area (4.96 billion ha) is used for agriculture and about 10% (1.37 billion ha) is arable land (FAO, 1999). Clearing new land for agricultureagriculture is increasingly being done on areas with no or little potential for sustainable agricultural production. Further increase in acreage is becoming less an option. Although technological advances will play an important role in feeding the world population, local food security will increasingly depend on the adaptive capacity of farmers. Adaptive capacity of human systems in Africa is low due to lack of economic resources and technology, and vulnerability high as a result of heavy reliance on rain-fed agricultureagriculture, frequent droughts and floods, and poverty (IPCC WG II, 2001).

Agricultural production is directly dependent on weather conditions, and will be affected therefore by climate change. Consequently this will have an impact on local food production. Temperature increases of 2–3°C will have a positive effect on production levels in temperate regions, whereas in most tropical regions the effect will be negative (IPCC WGII, 2001). Agriculture, being still the main source of income and livelihood for the larger part of the population in these regions, requires special attention.

Climate change can have an impact on agricultural sustainability in two interrelated ways: firstly, by reducing the long-term capacity of agroecosystems to provide food and fibre; and secondly, by inducing shifts in agro-ecological zones (Rosenzweig and Hillel, 1998). In rainfed agriculture in dry climates, dry periods, unfavourable for crop cultivation are of major concern (Rötter and van de Geijn, 1999).

In this chapter, the relations between climate change, notably its influence on drought, and food security for Sub-Saharan West Africa are briefly discussed. It is based on the results of the NRP financed project entitled: 'Impact of climate change on water availability, agricultureagriculture and food security in semi-arid regions, with special focus on West Africa' later renamed to 'Impact of Climate Change on Drylands' or ICCD, that brought together groups of scientists from an agro-biological, and climate modelling background with those from an anthropological, geographical and economic background. It also combined existing research efforts and expertise from a Wageningen-based consortium of natural scientists and economists ('Sustainable Land Use and Food

Security'), RIVM-Bilthoven and the national Graduate School for Resource Studies and Development CERES (for this project contributing geographical and anthropological expertise).

2. Climate change and food security

Availability and accessibility, in combination with the quality of food, determine whether an individual or group is food secure. Food supply is closely related to food production and food import levels. Accessibility, or the possibility to actually acquire the food products, is closely linked to socio-economic factors. The quality or nutritional value of these products is the result of the food production and processing chain, starting on the farmer's field and ending at the point where the consumer acquires and prepares the product for consumption.

Generally, food security depends on both availability of and accessibility to food. Even if food is available, it may not be accessible if people have no entitlements to food, that is, can access food based on an understanding that they have the right or means to obtain it. This can be through their own production, trade, employment or other resources available to the household, or through social relationships. An important entitlement is based on trade, when goods or services are traded and bartered or sold to obtain food products. In famine situations, however, entitlements to relief and international social security may play a crucial role in the survival of people (de Waal, 1989).

Taking an agronomic perspective, food security starts from crop production at the field level, moving via the farm and region to the national level; with the changing spatial scale, the focus of research changes from an agronomic or biophysical to a socio-economic and political analysis (Figure 7.1).

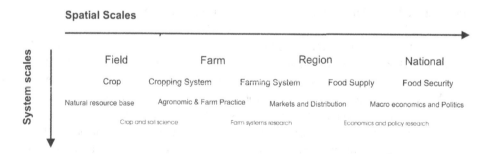

Figure 7.1. Spatial and system scales linking crop production to food security.

In rainfed agricultureagriculture in dry climates, dry periods unfavourable for crop cultivation are of major concern (Parry, *et al.*, 1988; Rötter and van de Geijn, 1999; Hulme, 2001). The performance of the rainy season is crucial as it largely determines whether there will be local food shortages or not. The Encyclopaedia Britannica gives the following definition of drought: 'it is the lack or insufficiency of rain for an extended period that causes a considerable hydrologic imbalance and, consequently, water shortages, crop damage, stream flow reduction and depletion of groundwater and soil moisture.' Drought is a normal, recurrent phenomenon that occurs in virtually all climatic zones, although its characteristics may vary significantly among zones (Hulme, 1995). Recent droughts in both developing and developed countries have underscored the vulnerability of all societies to this natural hazard, although some groups are more vulnerable than others are. The interaction between the natural event and the demand on the water supply determines the impact of a drought event.

Information on evapo-transpiration is crucial when quantifying the effect of drought stress on crop performance. Evapo-transpiration is a complex process, combining transpiration and evaporation. Transpiration is the process of water transport from the soil via the crop root system and its stems and leaves into the atmosphere. Evaporation is the upward transport of water from the soil surface to the atmosphere. Both processes are driven by the potential differences between the water in the soil and the atmosphere. The potential rate of evapo-transpiration is determined by the combined effect of the demand of the atmosphere and the properties of soil and crop.

Semi-arid regions are characterized by seasonal alternation of humid and arid periods where drought is a common phenomenon during the so-called dry season(s). Under the unimodal rainfall regime of sub-Saharan West Africa, the growing season for rainfed agricultureagriculture is restricted to the rainy season. More harmful are deviations from the seasonal rainfall patterns: unexpected drought, related to abnormal rainfall failure, which may occur in all climate zones. Such failures are predictable at a regional scale (Philippon and Fountaine, 1999; Thiaw *et al.*, 1999); at small scales the predictability reduces dramatically.

While the effect of rainfall amounts and their variability, whether annual, intra-annual or longer term, on production levels is important, there are a number of other characteristics in the sphere of the physical environment that influence yield levels and agricultural production: soils and their fertility in particular, but also geomorphological characteristics, competition with weeds, and pests and diseases to name the most important. Often the effects of various factors compound in affecting crop production in a negative way, e.g., pests and diseases being more threatening when plants are exposed to water stress.

3.   Food security and sustainable livelihood strategies

Using the concepts of vulnerability and sustainable livelihood strategies, the ICCD study tries to learn from the past to be able to cope with future situations, hence developing strategies for coping with climatic changes. Climate change is regarded an an additional stress to the system already marked by considerable inter-annual and intra-annual variability. The vulnerability of production systems and societies to climate change is related to the stability and resilience of these systems. Resilience refers to the ability of a system to absorb changes and restore production levels after a disturbing event. Farmers in unstable production environments have developed a range of livelihood strategies to cope with the inherent variability of the system and maintain food security.

Resilience in social sciences may be defined as the capacity of a person or group to anticipate, cope with, resist and recover from the impact of a disturbing event. People's response to risk is not only determined by expectations, but also by their values, preferences, culture, religion, etc. (Vogel, 1998). Responses of groups and mechanisms leading to social instability are extremely complex. Some of the short-term behavioural patterns of individuals and groups are predictable, but these are generally not the most interesting. Changes in long-term behavioural patterns are more important in understanding the consequences of decisions and pathways that individuals and groups may follow. These, however, are, even with perfect knowledge of all important factors, unpredictable. In low input, low-production environments, system dynamics, and hence farmers' response, will closely follow unpredictable external environmental factors.

To analyse the vulnerability of water and food security to climate change we need to address the existing capacities and existing resources, both natural and societal, that can be drawn on to deal with environmental changes. Also here the concepts: availability, accessibility and quality are essential. Technical, agronomic solutions can only address part of the problem; the same, however, holds for economic and socio-cultural solutions. Strategies aiming at security in unstable environments need to be integrative.

ICCD, recognising the complex interrelations among different types of capital (ecological, material, human and social) (Gladwin, 1997), developed a framework to analyse the various processes and functions related to sustainable food security, and to facilitate communication among the different disciplines. The layout of this framework is presented in Figure 7.2.

4. Biophysical responses to a changing climate

Agriculture is by far the most important economic activity in sub-Saharan West Africa (SSWA). It is the main source of income and livelihood in this region, as a large part of the population is active in subsistence agricultureagriculture. The arid and semi-arid regions in SSWA are among the most harsh and risky production environments in the world (van Keulen and Breman, 1990). Combined with a fast growing and mobile population, the pressure on both material and human capital will increase.

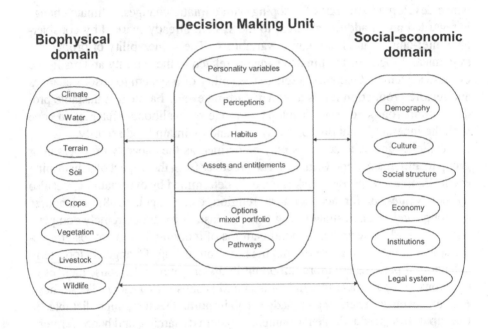

Figure 7.2.   Layout of the framework.

The rainy season is crucial for agricultural production, as rainfall during that period determines the level of food production. As indicated by GCM-scenarios (van den Born *et al.*, 2000), climate change may confront dryland West Africa with even lower and more variable precipitation, higher temperatures and higher potential evapotranspiration. This will result in higher risks for crop production, with the ultimate consequence of reduced food availability. The intrinsic low soil fertility does not allow continuous cropping without external inputs such as fertilizer (Breman, 1990). Labour availability is another constraint in crop production, as the continuous fight against weeds, pests and diseases is done mainly by hand. This section deals with questions related to climate change and climate variability in the region and their effect on crop production, focusing on rainfall, as that is the main driver of the agricultural production systems. The strong link between the arable and livestockinxxlivestock components and the role of urban centres in SSWA are not included in this analysis.

4.1   Drought Index

One way of quantifying drought is the use of a discrete index, serving as indicator for the degree of drought stress, which facilitates communication among scientists of different disciplines, planners and policy makers. Such indicators provide insight in the expected success rate for crop production in a given region and may trigger counter-measures by policy makers. Such drought indicators are static and lack predictive power.

Anticipated changes in precipitation, as calculated by GCMs have been used to calculate drought risk: Figure 7.3 clearly shows that these changes have a dramatic impact on the spatial distribution of the risk prone areas.

4.2　Modelling approach

Mechanistic dynamic crop growth models enable quantitative estimation of crop growth rates and yields under a variety of environmental and management conditions. Crop growth models are useful in understanding and exploring system responses to environmental conditions and management practices. Crop models may also be used for yield forecasting or can be applied in land-use evaluation, e.g., to assess production potentials of new cropping areas in thier dependence on climatic conditions and availability of water and fertilizer (Seligman, 1990).

The model used in this study is based on the LINTUL-type models (Bouman *et al.*, 1996), a simple generic crop growth model that simulates dry matter production on the basis of light interception and utilization, assuming constant light use efficiency. Potential evapo-transpiration determines the demand for water. For a closed crop canopy that completely covers the soil by leaves, this is in fact the transpiration demand. The amount of water needed to satisfy the transpiration demand can be calculated on the basis of crop-specific transpiration coefficients (Monteith, 1990).

The availability of water depends on rainfall and soil physical properties, with soil water between field capacity and wilting point assumed to be available for the crop. When soil water content falls below the soil texture-specific threshold, crop transpiration falls below the potential, proportionally reducing crop growth rate. The process of water movement into (rainfall) and out of (evapo-transpiration, and drainage) the soil profile is calculated with a capacity-type, dynamic soil-water-balance model. Total storage capacity is related to soil depth, set to a maximum of one meter. Because of the constant ratio between growth and transpiration, a constant transpiration coefficient is used.

The transpiration coefficient, under Sahelian conditions, may vary between 200 and 350; for cereals a value of 250 is commonly used. As this implies that 250 l of water is transpired per kg of crop dry matter produced, this coefficient is used to calculate the demand for water from the increase in biomass. When water supply can not meet the demand, growth rate is reduced proportionally. Crop yields follow from multiplication of the accumulated biomass over the growing period by the ratio of economic yield and total dry matter production, the harvest index, set to 0.43.

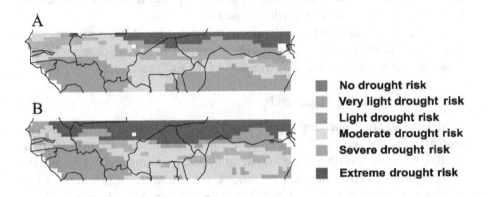

Figure 7.3. Changes in drought risk for 1990–2050, A depicts the current situation (based on 30-year average) B displays the projected drought risk for baseline A scenario, MPI (Max Planck Institut, Hamburg) -GCM.

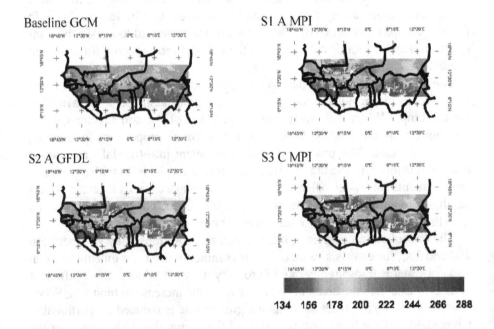

Figure 7.4. Start of the growing season resulting in highest water limited production level.

Calculations are performed using a daily time step for the baseline (1960–1990) and three climate change scenarios for Sub-Saharan West Africa. As a weather generator is used to downscale precipitation figures from monthly to daily values, calculations have been repeated 20 times for each soil-weather combination. As the start of the growing season varies, depending on the variable rainfall pattern, calculations have been started at two-week intervals from May 1st till the end of October. The resulting database, contains for each land pixel of 0.5×0.5 arcsec, calculated yields for 20 rainfall distributions for 12 starting points in the growing season.

Crucial for crop production is the onset of the growing season: due to 'false' starts crop failures may occur. Using the database, containing the simulated yields, the starting dates resulting in the highest production levels are extracted to establish the spatial variability in onset of the growing season. Figure 7.4 reveals, for all scenarios, an increase in variability of planting date.

Using the 20 calculated yield levels, for each soil-crop-weather combination we can, including some of the rainfall variability, asses the risk associated with a given management practice, such as planting date. For yield we concentrate on the probability that a given yield level is exceeded, rather than providing 'exact' yield levels. In Figure 7.5 the consequences of not adjusting the planting data are presented for the baseline and scenario (S2) A (Baseline A combines a medium population growth with a medium economic growth) GFDL (Global Fluid Dynamics Laboratory of Princeton University) scenario.

When planting around day 190 (early July), the highest yields are obtained for the baseline; when applying this planting date in the modified situation, a large part of the region, notably sandy soils and (semi)-arid parts, will experience higher risk levels, associated with lower average production levels.

Baseline   S2 A GFDL

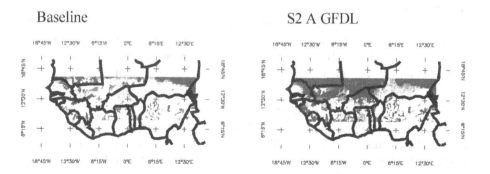

Figure 7.5. Probability of yield exceeding 2250 kg ha^{-1} for the 1960–1990 baseline and S2 A scenario using GFDL GCM, day of planting is 190. (black = 1, white = 0).

As indicated earlier, the results represent only a partial picture, as rainfall is but one of the biophysical factors affecting yield, while in addition socio-cultural, economic and political conditions are important. Input and output markets, institutional and legal frameworks and other similar contextual variables may have a considerable impact.

5.   People's responses to a changing environment

The local resource base and the contextual variables, all influence local decision-makers in different ways. Depending on their personal or household access to resources or their endowments in general, their entitlements and their (recent) history (their pathways), a selective process of valueing resources and their selective use, will result in very different levels of success in productivity and production. We concentrated on the question how climate change, notably changes in rainfall and rainfall variability, may affect yield levels. Using historical climate data, patterns of societal reactions to changing climate conditions were reconstructed, which however should be considered in the context of a changing society, of which demographic changes are probably the principal driving factors, next to changes in farm technology and farm management

Although many systems in nature may show seemingly unpredictable behaviour, they can also show remarkable levels of self-organization, just like a crowd of people. This capacity for self-organization presumably results from the incredibly complex feedback mechanisms among the objects making up the system. Because of the need for very strong feedback for realization of such self-organization, some of the most interesting examples of complex, adaptive systems are those involving human beings. Systems, including humans as their basic elements, can be the most unpredictable of all. Yet, since we have some memory of the past, and tend to make decisions based on our past success, it is possible that this feedback will result in recognizable patterns and hence predictability. Studying these patterns is a vital part of managing our society (Johnson, 1999).

Each individual or group has a particular 'portfolio of options'. Portfolios of theoretical, relevant and chosen options, as well as pathways, can be studied at the level of individuals (individual men and women, households, officials in institutions), but can also be applied to groups. For each individual or group we can make a history of portfolios of chosen options through time. This we call 'a pathway of chosen options'. It is plausible that decision making units that look similar (or belong to the same class of a typology) are all also characterized by similar pathways. Pathways arise out of the interaction between a Decision Making Unit and its environment. It refers to the specific path, which emerges out of the decisions taken by a specific decision-maker. The analysis of pathways implies that processes are analyzed.

Within the project nine portfolio options have been distinguished. These options or 'livelihood strategies' to achieve food security are briefly discussed, after which the pathways are touched upon.

5.1   Portfolio options

Storage of food   Storage of food, produced in years with relatively abundant yields, is an age-old practice, but it can also be a new or reinvented strategy, either at the level of individuals/households or at the level of communities or even as a result of national policies. Improved storage technology and reduced losses during harvest, during food preparation and during consumption may result in major improvements, because of the magnitude of these losses. Many 'coping strategies' during problematic times can also be based on investing in various types of assets in favourable periods and selling those in dire times. Assets can be in the form of accumulated wealth (gold, or other jewellery; animals; bank savings; land; house; household utensils) or in the form of productive capital (e.g., agricultural inputs, agricultural and non-agricultural machinery). In semi-arid and arid West Africa, the sale of small ruminants (goats, sheep) is a widespread 'first aid' strategy in case of emergencies (Slingerland, 2000).

Increasing food production   This may be attained by extending the food crop area into 'wastelands', forest areas, game- or bio-reserves or natural pastures and rangelands; the possibilities depend on the positions of relative strength of competing land users: arable farmers versus charcoal burners, game wardens, forest departments, pastoralists, etc., as well as 'pioneer' cultivators competing among themselves, often with an ethnic element in claims and counter-claims.
Alternatives could be:

- more efficiently using a variety of geographical niches;

- limiting the fallow period and – to achieve that in a sustainable way – maintaining or improving nutrient stocks by adding manure, fertiliser, compost, or by including crops, in a rotation i.e., legumes, that add nutrients to the system, etc.;

- improving soil and water management, including application of soil and water conservation measures; the scope for such practices partly depends on long-term security of access to these improved lands, and this often depends on indigenous, religious and/or state-defined land, water and other natural resource rights;

- adopting crops and varieties with higher or more secure yields per ha.; and by improving access to 'improved' seeds of these crops and varieties;

- by developing irrigation facilities, either using river water, canal or piped water or groundwater reserves; this may also increase the cropping intensity, i.e., the number of crops grown within a year;

- intensifying animal husbandry practices (e.g., improved veterinary care, higher-yielding species, stall feeding, improved husbandry); however, the dangers of intensifying resource depletion should not be overlooked (Savadogo, 2000; Slingerland, 2000); an alternative would be sending part of the animals to places that are better endowed with feed and water, if available;

- investing more labour and care in food crop cultivation, and in animal husbandry practices;

- investing more labour and care in gathering 'wild food', and/or accepting a diet that under 'normal' circumstances is regarded as inferior ('poor men's food', or 'hunger food') or taboo.

- Land use models, which include differences in soil resources, technologies /management, and farming systems, can be used to explore the consequences of various food production scenarios, and estimate their requirements in terms of inputs (Savadogo, 2000). In combination with results of crop growth models, such land-use analysis models can also be used to examine the consequences of climate change. Bio-economic simulation models can also shed light on the driving forces of changes in land use from a market and state intervention perspective (Brons et al., 1998; Ruben et al., 2001).

Intensification of food production   It might also be possible to indirectly invest in food production intensification, producing crops, raising livestock or collecting tree products for an external market and buying or bartering food in return. Possibilities depend on net terms of trade for the actual producers. Commodities, which have provided this opportunity for at least part of the recent period, have been cotton (for the sub-humid zone), groundnuts (for the semi-arid zone), charcoal (for both) and meat. The scope for these practices depends on price developments at the world market and, at regional markets for export products, on the price developments of food in these external markets, and on the marketing infrastructure, and transaction costs (including taxes, bribes, monopoly profits, etc.) in the areas of demand. This option only 'works', when for the producers/consumers the net impact of producing for the market results in higher food availability than if they would have produced the food themselves. Senegal, for example, has for quite some time clearly opted for a groundnuts-for-cereals exchange, importing one third of its food requirements during the 1990s. International and national market dynamics and government policy all play a role in determining likely outcomes, with after the early 1990s a shift away from policy-driven adjustments to market-driven adjustments (de Haan *et al.*, 1995; Sanders *et al.*, 1996.).

Marketing non-agricultural products This would imply changing a basically agricultural economy to one in which other activities, such as mining or producing handicrafts or industrial products becomes more important and where non-agricultural producers get higher net rewards for their labour time than if they would have produced their own food. The scope for this option very much depends on availability of niches in the world market (e.g., 'ethnic art'), or on making use of comparative advantages of the region, e.g., low wages, low environmental sink costs, or absence of effective anti-pollution policies. A breakthrough of Asian 'tiger'-type industrialisation is far away, though, given the extremely low education level in the area: primary net enrolment figures of below 50% in Senegal to below 20% in Mali, and secondary enrolment below 10% for men and below 5% for women in most areas (World Bank, 1995).

Selling services This could be attained by attracting tourists and/or other visitors from abroad, either to enjoy nature and landscape adventurism ('ecotourism'), culture (either in its pure 'ethnic' form or as a hybrid popular culture e.g., west-African popular music), or sexual services and the 'image of care'.

Selling labour If the scale of analysis is the individual decision making unit, all income acquired by selling labour for wages (permanent, seasonal, casual; for government agencies, Non Governmental Organisations or the private sector; near home or far way) is relevant. If the scale of analysis is the region, the relevant option is the acquisition of remittances, i.e., food, money or material goods, either regularly or irregularly, sent by people originating from the region, but working elsewhere, either on 'hunger trips', on seasonal trips or for extended periods. Remittance networks may link the area to many parts of the globe, although mostly the links are rather bilateral. In dryland west Africa, the links with various parts of the (urban) coastal zone have become extensive, but many Sahelians have now taken up residence in less densely populated humid and sub-humid areas towards the south, but still far from the coast. A growing number of Sahelians has migrated to Europe, and contribute considerable sums of money to the region. Figures for 1990 show a total remittance sum (net transfers at the national level) of Burkina Faso and Mali together of 557M$, while total export income was 769M$ and total official development aid 800M$ (World Bank, 1992). In recent years, the net transfers have become less important, though, because of a deepening crisis in Ivory Coast, which was the major destination for migrants from the north. In 1998, total net transfers of remittances for these two countries were 313M$, while total export income was 1036M$ and total official development aid 754M$ (World Bank, 2000).

Social security arrangements   Social security arrangements are either directly, in the form of social care or in the form of food aid by local governments or non-governmental agencies or by foreign donors through development assistance or by international non-governmental organisations, or indirectly in the form of income support, for which food can be bought: pensions, social security payments, bank or trader's/moneylender's credit.

Food aid (mostly cereals) has been a structural phenomenon in Sahelian countries since the famine of 1973–'74. Part of cereal imports comprises food aid (in 1993 40% in Mali, 25% in Burkina Faso, 20% in Ghana and 12% in Senegal), but in most years it only covers a minor part of all food needs (in 1993 less than 5%; World Bank, 1995). However, food aid might be there to stay as a structural phenomenon, as development assistance has already become a structural phenomenon in most study areas, with official development aid between 8 and 25% of Gross National Income and a major element of government budgets (World Bank, 1997).

Stealing   This can either be directly (stealing food from other peoples' fields or storage places, or by stealing/raiding animals) or indirectly by stealing material goods, and exchanging those for food, or by demanding bribes, contributions, etc. At regional scale, waging a war or raiding neighbours and transporting the 'booty' back home is a possibility.

Reducing food demand   This can also be a 'solution' to a pressing or threatening food problem: at the level of scale of individual decision making units, it may form a strategy to accept lower food intake; eating fewer meals a day or eating food of lower quality; at the level of households, a strategy may be to 'outplace' members (children mostly), e.g., bring them under the care of relatively well-to-do family members or to 'marry off' daughters; a strategy may also be to break up (joint) families and accept responsibility for a lower number of household members.

At the level of scale of regions it can be acquired by:

Out-migration   does not necessarily results in remittances and it should not only be judged by its capacity to generate remittance income. Out-migration can also reduce the burden on those remaining, and reduce the overall food needs.

Reducing natural population growth   through lower birth rates. Family planning is regarded in some circles as the most obvious measure to reduce population growth (Breman, 1998), which is a result of very high total fertility rates and improved health conditions; birth rates can also decline as a result of severe physical stress, due to malnutrition, resulting in higher abortion rates and/or infertility;

Declining natural population growth through higher death rates. Gross death rates are still high by world standards, but much lower than in the 1960s and they are expected to further decline, despite the impact of diseases, such as AIDS. What could happen though is a chaotic explosion of violence, as elsewhere in Africa, and that of course could have a major impact on food demand. As violent disruption of livelihoods will also result in disruption of agricultural production, land management and distribution channels and in a destruction of food and capital stocks (including existing investments in natural resources and infrastructure), it will also have a negative impact on local food supply, with long-term effects that may completely undermine what might be cynically called 'gains' of lower food demand.

5.2 Pathways

The study of pathways is essentially about processes of decision-making. By following the development of a Decision Making Unit over time, by identifying the key decisions or chosen options that have led to its present state, it is possible to gain insight into the interactions between the decision maker and their environment and will help in making forecasts on future decisions and pathways.

For example Decision Making Units that have survived a terrible drought are more likely to respond differently to a good year than people who didnot experience the drought. Likewise, younger generations will have a different attitude to farming and herding and life in general than older generations. Although living in the same time and environment their perceptions are different, this will influence their preferences and most likely also their goals and the decisions they take.

Pathways is oriented towards the dynamics of decision-making processes, i.e., to pinpoint under what opportunities and constraints which type of actors and groups of actors are likely to follow specific pathways to mitigate instability. This pinpointing may result in the formulation of a number of 'rules of the game' which people use to determine their pathway. This qualitative information may be fed into the formal decision-making models.

6. Concluding remarks

Scenario analyses, on the basis of different global circulation models, show a wide variety of outcomes for 2050, but there seems to be consensus that most of dryland West Africa is likely to become appreciably drier (higher temperatures (lower air humidity levels) and lower rainfall levels). The consequences of these projections are an increase in high-risk environments for agricultureagriculture, and a further southward shift of the arid and semi-arid zones. Changes in rainfall distribution will aggravate that situation and lead to additional stress on agricultural production in these areas. Simulation studies clearly reveal a delay in the onset of the growing season and associated lower yield levels. The very unfavourable initial (biophysical and economic) conditions in combination with the low adaptive capacity are major bottlenecks in trying to alleviate food insecurity.

Droughts should be recognised as part of a highly variable climate, rather than treating them as natural disasters. Current management practices should be adapted to capture this concept. Farmers should be provided with climatic information as a basis for priority setting and flexible management decisions. Drought or disaster relief should be the last option drought management or adaptive management geared to the prevailing climate variations, including droughts, should be the main focus.

In accordance with observations in other developing regions in the world, in West Africa's drylands the importance of non-agricultural sources of rural livelihood strongly increases, both locally, and (mainly) as remittances from work in other sectors.

The agricultural sector is restructuring, both because of changing environmental conditions, and as a result of increasing urbanisation. Increased demand for food, water and energy provides opportunities for farmers in the surroundings of (big) cities, but also for farmers at rather large distances (charcoal and non-perishable food coming from areas hundreds of kilometres away). Another development in the rural areas is the increase in the contribution of cash income from more remunerative elements of agricultural production: cotton, vegetables, meat, milk and of charcoal.

In general, all over dryland West Africa, but in the north even more than in the south, there was a strategic attempt by many farmers to develop a multi-locational, and multi-sectoral household economy, both in agricultureagriculture and outside the sector.

Traditional socio-cultural dichotomies in livelihoods gradually disappear, but cultural and ethnic identities and identification processes are still important (Slingerland, 2000), and retain political impact at the national and local levels; the decentralisation processes of government power and the growing influence of non-governmental agencies can be expected to have a major impact on the culturally diverse ways of coping with adverse situations and on access to natural resources and to livelihood options in general.

The rather positive findings may well be based on the rather favourable rainfall situation of the 1990s and on the lessons that people have learned from the droughts in the 1970s and 1980s (individually and as collectives, assisted by foreign donor agencies that are quite important in West Africa). If the 1990s prove to be a temporary upswing in a climatological cycle that is tilted downwards (which we expect), government and donors are advised to be more cautious and to prepare for more difficult periods to come. Research efforts are needed to support this advice.

References

Bouman, B.A.M., and van Keulen, H. and van Laar, H.H. and Rabbinge, R. (1996) The 'School of de Wit' crop growth simulation models: A pedigree and historical overview. Agric. Syst. 52, 171–198.

Breman, H. (1990) No sustainability without external inputs. In: Sub-Saharan Africa; Beyond adjustment. Africa seminar, Min. of Foreign Affairs, DGIS, 124–134.

Breman, H. (1998) L'intensification agricole au Sahel: vouloir c'est pouvoir. In: Breman, H. and Sissoko, K. (eds.) L'intensification agricole au Sahel. Karthala, Paris, 23–34.

Brons, J.E. and Tour, M.S.M. and Oudraogo, B.S. and Ruben, R. (1998) Driving forces of changes in land use in Burkina Faso and Mali. Supply-response analysis for cereal production. Wageningen WAU, Ouagadougou CEDRES and Bamako CPS. ICCD paper.

De Waal, A (1989) Famine that kills: Darfur, Sudan, 1984–1985. Clarendon Press, Oxford

FAO (1999) FAOSTAT Agriculture Data. http://apps.fao.org

Gladwin, T.N (1997) A call for sustainable development. Financial Times Limited. http://www.bus.umich.edu/ft/gladwin.pdf.

Haan, L. de and Klaasse Bos, A. and Lutz, C. (1995). Regional food trade and policy in West Africa in relation to structural adjustment. In: Simon, D. and van Spengen, W. and Dixon, C. and Nrman, A. (eds): Structurally adjusted Africa - Poverty, debt and basic needs. London: Pluto Press.

Hulme, M. (1995) Climatic Trends and Drought Risk Analysis in Sub-Saharan Africa. University of East Anglia, Climatic Research Unit, Norwich.

Hulme, M. (2001) Climate perspectives on Sahelian desiccation: 1973-1998.Global Environmental Change 11, 19–29.

IPCC WGII (2001) Summary for Policymakers. Climate Change 2001: Impacts, adaptation, and vulnerability. http:www.ipcc.ch.

Johnson, N. (1999) Arrows of time. The Royal Institution Christmas Lectures, London.

Monteith, J.L. (1990) Conservative behavior in the response of crops to water and light. In: Rabbinge, R. and Goudriaan, J. and Keulen, H. van Penning de Vries, F.W.T. and Laar, H.H van (eds.) Theoretical production ecology: reflections and prospects, Simulation Monograph 34, Pudoc, Wageningen, 3–16.

Parry, M., and Carter, T.R. and Konijn, N.T. (1988) The impact of climatic variations on agriculture: Vol.2. Asssessment in semi-arid regions, Kluwer Academic Publishers, Dordrecht, The Netherlands.

Philippon, N and Fountaine, B (1999) A new statistical predictability scheme for July-September Sahel rainfall (1968 - 1994). Comptes Rendu de l'Academie des Science. Serie II Fasicule A-Sciences de la terre et des plantes, 329, 1–6.

Rosenzweig, C. and Hillel, D. (1998) Climate Change and the Global Harvest. Potential Impacts of the greenhouse effect on Agriculture. Oxford University Press.

Rötter, R. and van de Geijn, S.C. (1999). Climate Change effects on plant growth, crop yield and livestock. Climate Change 43, 651–681.

Ruben, R. and Berg, M. and van den and Shuhao, T. (2001). Land rights, farmers' investments, and sustainable land use: Modelling approaches and empirical evidence. In: Heerink, N.and van Keulen, H. and Kuiper, M. (eds.) Economic policy and sustainable land use. Recent advances in quantitative analysis for developing countries. Physica Verlag, Heidelberg, New York, 317–334.

Sanders, J.H. and Shapiro, B.I. and Ramaswamy, S. (1996) The economics of agricultural technology in semi-arid sub-Saharan Africa. The Johns Hopkins University Press, Baltimore.

Savadogo, M. (2000) Crop residue management in relation to sustainable land use. A case study in Burkina Faso. Ph. D. Thesis, Wageningen University. The Netherlands.

Seligman, N.G. (1990) The crop model record: promise or poor show? In: Rabbinge, R. and Goudriaan, J. and van Keulen, H. and van Laar, H.H. and Penning de Vries, F.W.T. (eds.) Theoretical Production Ecology: reflections and prospects. Sim. Monogr. 34, Pudoc, Wageningen, 249–263.

Slingerland, A.M. (2000) Mixed farming: Scope and constraints in West African Savanna. Ph.D. Thesis, Wageningen University, The Netherlands.

Thiaw, W.M. and Barnston A.G. and Kumar, V. (1999) Predictions of African rainfall on the seasonal timescale. Journal of Geophysical Research - Atmospheres 104 (D24), 31589–31597.

van den Born, G.J. and Schaeffer, M. and Leemans, R. (2000) Climate scenarios for the semi-arid and sub-humid regions: a comparison of climate scenarios for the dryland regions in West Africa, from 1990-2050, Wageningen: NOP report no: 410 200 050.

van Keulen, H. and Breman, H. (1990). Agricultural development in the West African Sahelian region: a cure against land hunger? Agric., Ecosyst., Env. 32, 177–197.

Vogel, C.H. (1998) Vulnerability and Global Environmental Change. LUCC Newsletter 3 Special Issue: The Earth's Changing Land Conference.

World Bank (1992, 1995, 1997, 2000) World Development Reports, Washington. Oxford University Press.

Chapter 8

LAND-USE CHANGES INDUCED BY INCREASED USE OF RENEWABLE ENERGY SOURCES

S. Nonhebel
Center for Energy and Environmental Studies, Groningen

1.   Introduction

In the previous chapters of this book attention is paid to the impact of land use on global change. Changes in land use studied concern changes from natural vegetation into an agricultural production system (and sometimes the other way round, from agricultural into forest again) and changes within the production systems due to change in management.

The land-use change only accounts for a part of the global change. The observed increase of the atmospheric CO_2 concentration is mainly due to the increased use of fossil fuels since the industrial revolution. Reduction of the global CO_2 emissions related to energy use is essential to mitigate climate change. This reduction can be achieved via two routes: improvement of the energy use efficiency on one hand and the increased use of non-CO_2 energy sources on the other. In the last decades a large number of energy saving programs were conducted both in industry and in households. These programs focus on increased use of energy efficient appliances and/or improved insulation of buildings etc. On the national level an increase of energy use efficiency has been observed.

On the other hand attention is paid to introduction of non-CO_2 energy sources, such as electricity from windmills and photovoltaic cells, energy from biomass and use of solar heating systems. Several types of subsidies exist both for industry and for households to increase use of renewables. The Dutch government aims at a 10% share for renewable energy sources by the year 2020 (Ministerie van Economische Zaken, 1995), and for the more distant future, larger shares are needed to prevent global change in a sustainable society where all energy originates from renewables.

A.J. Dolman et al. (eds.), Global Environmental Change and Land Use, 187-202.

This increased interest in renewable energy sources is the link with land-use changes evaluated in the previous chapters. Renewable energy sources such as photovoltaic cells, biomass and solar heating systems obtain their energy directly from solar radiation. The amount of energy that can be generated with these systems depends on the surface that can be used for the interception of solar radiation. Presently only 1% of the Netherland's energy demand is fulfilled with renewable energy. The required increase of renewable energy share (1–10%) implies that more surface is required to intercept solar radiation. This chapter focuses on land-use changes as a result of changes in the energy production system. A short description of the energy system is given and attention is paid to quantities of energy used and to the various energy carriers utilised in the energy system. The characteristics of photovoltaic cells and biomass as energy sources are given with respect to their impacts on land use. The land claim generated by an energy system that includes a large share of renewables is determined and the consequences of this land-use change for global change are discussed.

2.   The past and present of the energy system

Figure 8.1 shows the development of global energy use since the mid-19th century; both the total consumption and the main energy sources are shown.

Until 1900 the consumption was more or less constant at 20 EJ yr^{-1}. After the industrial revolution the total energy use increased to 400 EJ yr^{-1}. Besides this enormous increase in total consumption, also a change in energy sources is evident. Before the industrial revolution energy consumption was mainly supplied by wood and in present days it is a mix of several energy sources. These changes in energy sources (Figure 8.2) came along with large changes in society.

Before the industrial revolution the total consumption was low and energy conversion was limited to burning of fuel wood for heating and cooking. Humans, horses, windmills and watermills provided mechanical power.

The introduction of the steam engine powered by coal caused a large change in the energy system. It was the first conversion from fossil energy into work. Further, since coal can be transported and stored, the work could be done on the site where it was required. This in contrast to windmills and watermills where the work had to be done on the site where the energy was. This led to a new organisation of production: the factories. The development of the mobile steam engine had large impacts for the transportation system. At the beginning of the 20th century coal had replaced nearly all traditional energy sources and global energy use had increased to 50 EJ yr^{-1} (Figures 8.1 and 8.2).

The introduction of electricity as energy carrier (that can supply light, heat and work) and the development of the internal combustion engine triggered the

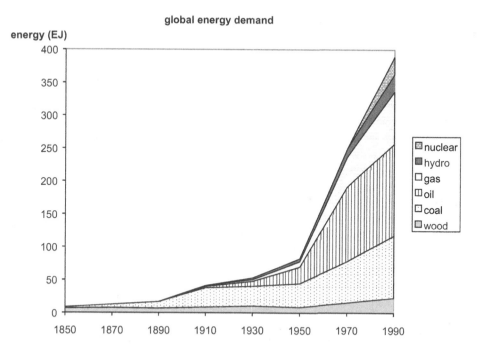

Figure 8.1. Global energy consumption (EJ yr^{-1}) and the energy carriers used (adapted from Grubler and Nakicenovic, 1997).

second transition: the introduction of oil as a carrier and later on the introduction of natural gas. At the end of the 20th century total energy use is increased to 400 EJ yr^{-1} and this amount is supplied by a variety of sources: natural gas, oil, coal, hydropower, nuclear power and wood.

The changes from 1850 onwards showed an increase in total use and a change in sources. These changes in energy sources came together with technological, economic and institutional changes in societies and these changes are interrelated. The situation in individual countries deviates from the global average. Differences are found in the total amount of energy used, the energy carriers in use and the timing of the change from one carrier to the other.

Major differences can be observed between the industrialised countries and the developing countries. Industrialised countries use 200 GJ capita^{-1} yr^{-1}, while in the rest of the world it is only 35 GJ capita^{-1} y^{-1} (Hall *et al.*, 1993). Next to the total amount the carriers used differ. In the developing countries biomass (wood), is still an important energy source (38%), while in the industrialised world it only accounts for 3% of the energy supply (Figure 8.3). With respect to the use of hydropower, no difference is found; in both the industri-

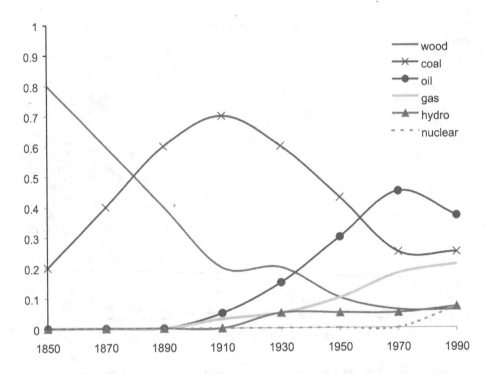

Figure 8.2. The share of the various energy carriers in the total energy supply used (adapted from (Grubler and Nakicenovic, 1997).

alised countries and the developing countries, 5% of the energy is generated with hydropower. However, it should be realised that the renewable energy use per person is similar all over the world (8% of 200 GJ= 16 GJ and 43% of 35 GJ= 15 GJ).

Between industrialised countries also large differences exist. These differences mainly include the energy sources used not much difference in the total energy use per capita is found. The types of energy sources used in the energy systems of individual countries depend on the availability of natural resources.

In the Netherlands, in contrast to the global picture, wood has never been of importance in energy supply. In the pre-industrial economy peat was the major energy source (Figure 8.4). In 1850 already 50% of the energy was obtained from coal and in 1910 nearly all energy originated from this source. The exploitation of the large natural gas reserves in the 1960s led to a sharp decline of the use of oil. At the end of the 20th century natural gas became the most important energy carrier. The contribution of the renewable energy sources to the total is limited to 1%. This is in contrast with for instance Sweden where the electricity from hydropower accounts for 20% of the total energy use.

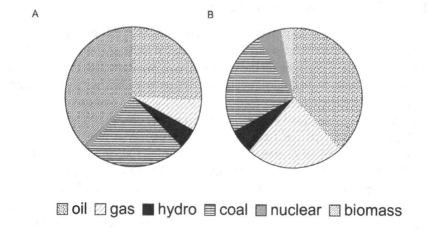

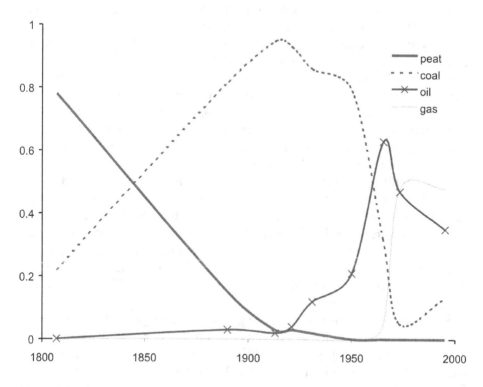

Figure 8.3. The share of energy carriers to the total energy supply: comparison between developing (A) and industrialised countries (B).

Figure 8.4. Contribution of various energy carriers to the total energy demand in The Netherlands.

2.1   Present use of solar radiation

The total incoming solar radiation mounts up to $3.5\,10^6$ EJ annually. On earth solar radiation is converted via physical or chemical routes. The physical routes include temperature changes on the earth surface, pressure differences and accompanying winds, hydrological cycles etc. Therefore electricity generated by windmills and hydropower is considered indirect use of solar radiation.

The chemical route runs via photosynthesis where electromagnetic energy is converted into chemical bindings. A part of this energy is used for respiration processes, the remainder is converted into plant material: the primary production. The terrestrial Net Primary Production (NPP) is estimated to be 2800 EJ yr^{-1} (Wright, 1990). The chemical route interferes with the physical routes, since photosynthesis in plants requires certain climatological conditions, such as the availability of water. Further, most crops show preferences with respect to the prevailing air temperatures.

An important use of solar radiation involves the production of food. In agricultural systems crops are used to intercept solar radiation and the products of the photosynthesis process are converted into human edible materials. Per person, 10 MJ day^{-1} of food energy is required. This is 3.6 GJ $person^{-1}$ annually and with a population of six billion people on earth this mounts up to 22 EJ yr^{-1}. This is 1% of the global annual NPP, which seems limited, however the impact of human food systems on the planet is far larger. The food energy consumed partly originates from livestock. The conversion from plant material into animal products is poor. For the production of 1 MJ of meat about 8 MJ of plant material is required. Further, humans tend to eat only a part of the crop (only the grains of a wheat plant and the apples of the tree). This implies that far more plant material has to be produced than the 10 MJ per person per day. (Wright, 1990) even estimated that for the human food production 20–30% of the NPP (650 EJ) is diverted from its flow through natural ecosystems. Which indicates the enormous impact of the production of food on global ecosystems.

It should be realised that the presently used fossil fuels also originate from the chemical solar conversion route. Oil, coal and natural gas originate from plant material produced 300 million years ago. The combustion of fossil fuels is the use of ancient solar radiation and the emitted CO_2 is the CO_2 incorporated in the plant material long ago.

So the food consumed and the energy used both originate from the same system. The difference in magnitude used is striking: a person in the industrialised world uses 3.6 GJ as food and 200 GJ as fuel annually. In the developing countries the difference is smaller but still includes a factor of 10: 35 GJ for fuel and 3.6 GJ for food.

2.2   Solar energy as a renewable energy source in a modern society

Since solar radiation is the largest energy source on earth, the search for non-CO_2 energy sources is focused on the use of incoming solar radiation. The incoming solar radiation on earth is huge (3.5 10^6 EJ) compared with the fossil energy used (400 EJ). The present total annual energy consumption in the world is only 0.01% of the incoming solar radiation on the planet.

The situation in The Netherlands is less favourable: fossil fuel use is 2% (2.3 EJ) of the incoming solar radiation (110 EJ), but still the incoming energy is far larger than the consumption. So there is a potential for using solar energy as an energy source, even in a small, densely populated and cloudy country as The Netherlands.

However, the characteristics of the incoming solar radiation (low density and large variation) are such that the options for direct use in the present energy system are limited. The present applications of direct solar energy involve the generation of heat as in solar water heating systems, glasshouses and sun lounges. Modern societies, however, require large quantities of electricity and transport fuels. To be of use for present modern societies, solar radiation has to be converted into a carrier that fits in the present electricity and transport systems.

For the more distant future other options are possible. In the previous paragraph it was shown that energy carriers used in societies change in time. A change from one carrier to another involves technical, economical and institutional changes (i.e., construction of a natural gas grid). All these changes require time, but are not unlikely. This implies that for the short term (as the 10% renewable in 2020) carriers have to fit in the present system, but that for the longer term carriers that do not fit in the present system might be an option since energy systems can adopt to new situations.

The conversion can occur again via the physical route and the chemical one. The physical route includes systems like windmills, where occurring wind energy is converted into electricity but also the photovoltaic systems. The chemical conversion route is via photosynthesis. In all conversion processes energy is lost. Here two conversion routes are compared: a physical one (photovoltaic cells) and a chemical conversion via photosynthesis (energy from biomass).

Energy from photovoltaic cells (PV-cells)   In PV-cells (or solar cells) solar energy is converted into electricity. In principle a PV-cell consists of two layers; the photoelectric effect of solar radiation (photons) creates a difference in electric potential between the front and back layers of the cell. This difference is used to generate a direct current. This can be converted in alternating current with the aid of a transformer. Several types of solar cells exist; for an overview see (Green, 1993). Solar cells are combined to form solar modules. Solar modules can be placed on roofs of houses and other buildings and connected to the grid, but stand-alone systems also exist. The efficiency of the systems varies between 5–30%, depending on the type of system used and prevailing circumstances. In this paper an efficiency of 15% is used for the calculations, since that is the highest efficiency for presently available systems. The price of electricity obtained from solar cells is 135–150 ct per kWh (Ministerie van Economische Zaken, 1999; KPMG, 1999).

Energy from plant material In photosynthesis, solar energy is converted into chemical energy (glucose). The assimilation products of the plant (gross primary production) are used for maintenance of the existing plant material and for the production of structural plant material (cellulose, proteins, fats, etc: the net primary production). This structural plant material can be used for various purposes. In general it is used as food or feed, but it can also be used as an energy source. The use of plant material as an energy source leads to the emission of CO_2. This CO_2, however, has been taken up from the atmosphere and incorporated into structural plant material by the plant in the preceding growing season(s). The net emission of CO_2 from this energy source is therefore zero. Substitution of fossil fuels with energy from biomass therefore leads to a reduction of CO_2 emissions.

The use of biomass as an energy source is not new, since wood was the most important energy carrier in the past. And as mentioned earlier, in developing countries still 40% of the energy used is obtained from biomass. However, until now biomass has been mainly used for heating and cooking purposes. The present western world societies require other carriers and to be of use in a modern society plant material has to be converted. The most frequently mentioned routes are conversion into a transport fuel (biodiesel or ethanol) or into electricity. These conversion processes are not free of costs and require energy themselves. In a comparative study of suitability of plant material for energy (Lysen *et al.*, 1992) it was shown that use of wood from short rotation forestry systems as a feedstock for electricity generation was the most efficient method (from the energy perspective) to obtain energy from plant material.

In short rotation forestry systems, fast-growing trees like willow and poplar are planted at a high density (one tree m^{-2}). After a couple of years (four to seven) the crop is harvested by coppicing. The stubs re-sprout and a few years later the crop can be harvested again. It is estimated that such a system can remain productive for about 20 years.

The highest solar energy use efficiency is obtained when high input systems are used (Nonhebel, 2002). These systems include the use of heavy machinery, use of fertilisers and pesticides. This type of production system can only be realised on high quality soils (comparable to soils required for food production). The wood chips are combusted in a conventional coal plant.

The solar energy use efficiency of these short rotation systems is estimated to be 0.5–1%. (The potential yields are 10–20 ton ha^{-1} year^{-1} (Nonhebel, 1997), heating value of wood is 18 MJ kg^{-1}, annual incoming radiation in The Netherlands is 3500 MJ m^{-2} yr^{-1}).

The use of fertilizers, pesticides and the frequent harvesting makes this production system comparable with high input agricultural systems rather than with natural forests. It is possible to grow biomass under low input conditions (no fertilizers etc.) but than yields drop to about 2 ton ha^{-1} yr^{-1} and become comparable to NPP in natural forests. Such low yields, however, are uninteresting from the energy perspective (Nonhebel, 2002). The costs for wood chips as feedstock for electricity plants are estimated at 12 ct per kWh, while feedstock costs for coal are 4 ct per kWh (Ybema *et al.*, 1999). (Electricity is sold to households for about 25 ct per kWh, excluding taxes).

3.   Land requirements for energy generation purposes

In the systems described above solar energy is intercepted and converted into electricity or into plant material. The quantity of energy that can be generated with these systems depends on their solar energy use efficiencies and on the area that is available to intercept the solar radiation. Based on the information given in the previous paragraphs the land requirements for energy generation were calculated.

In Table 8.1 the results of these calculations are given both for the national and for the global situation. When all energy used in The Netherlands (2.3 EJ) has to be fulfilled with the aid of PV-cells, about 0.43 Mha is required, which equals 13% of the Dutch land area. When this has to be done with biomass this would require 6.5 Mha, which is nearly two times the total Dutch land surface. The potential contribution of biomass to the energy supply in The Netherlands is therefore rather limited. The use of PV-cells for energy supply is more realistic. A major advantage of PV-cells is that roofs and fronts of houses and other buildings can be used to intercept this radiation. It is estimated that about 400 km^2 (= 0.04 Mha) of roofs and other surfaces in The Netherlands can be used for energy generation with PV systems (KPMG, 1999).

On a global scale a different picture is obtained: 11% of the global land surface is needed for the production of energy when a biomass system is used. This value is in accordance with other studies to potentials of biomass as energy source on global scale (van den Broek, 2000). So in principle energy obtained from biomass systems can provide the present worldwide energy requirements.

Table 8.1. Determination of surface requirements for energy production using PV-cells or biomass production systems.

System	The Netherlands		World	
	biomass	*pv-cells*	*biomass*	*pv-cells*
Incoming rad. (MJ m^{-2})	3500	3500	5200	5200
Energy yield (GJ ha^{-1})	350	5250	520	7800
Energy use (EJ)	2.3	2.3	400	400
Area required	6.5 Mha	0.5 Mha	0.8 Gha	0.05 Gha
Total land area	3.5 Mha	3.5 Mha	13 Gha	13 Gha
Percentage total land area	200	13	6	0.4

4.   Potentials of PV and biomass in near future

4.1   Photovoltaic systems

The calculations in the previous paragraphs show that in the long run energy from PV systems has the largest potential to fulfil national needs, for energy. In the short-term (20 years) factors other than just the systems efficiency play a role. First of all the price per kWh is important. As mentioned before electricity from solar cells is far more expensive than electricity from fossil fuels (150 versus 25 ct per kWh). These high costs will hamper the introduction of these systems anyhow. However, even when this obstacle is removed, for instance through large investments by government, other problems will arise.

These problems have to do with the implementation in the present energy system. In order to be useful in a modern society electricity has to be available on demand (time and place). At a national level electricity demand varies between 5.000–10.000 MW. It shows fluctuations over the day (during the night demand is smaller) and differences between seasons. Figure 8.5 shows typical summer and winter demand patterns. In summer evenings demand is smaller since no electricity is used for lighting. The diurnal variation is larger than the seasonal differences. In a fossil fuel (stock) driven society it is possible to adjust the supply to the demand since one can determine the magnitude of the flow through switching electricity plants on and off. When electricity is obtained directly from the solar radiation flux this is no longer the case and buffers are needed.

Figure 8.5 also shows the electricity supply when 60.000 ha of PV cells are installed. This 60.000 ha is the area of PV required to fulfil the present annual electricity demand (Table 8.2).

The supply is determined by multiplying the PV-cell surface with the average hourly incoming radiation per square meter (Meteorological station de Bilt, (Velds, 1992)) and the efficiency of the PV-cells. Incoming solar radiation shows an enormous variation both over the day and among seasons; the amount of electricity obtained from the PV-cells follows this pattern. The incoming

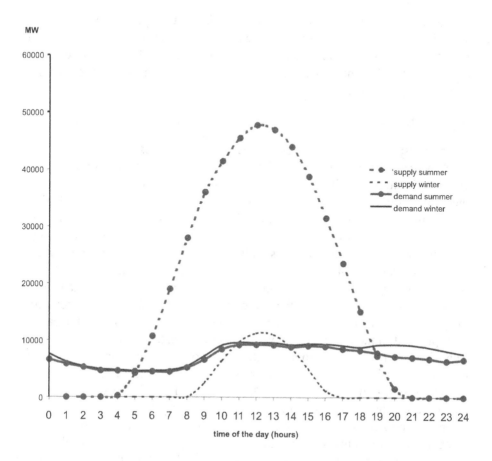

Figure 8.5. The Dutch national energy demand in summer and winter and the supply in these seasons when electricity is generated with PV systems.

Table 8.2. Determination of the acreage of PV-cells required fulfilling the annual electricity demand in The Netherlands.

National electricity demand (PJ yr^{-1})	330
Average incoming solar radiation MJm^{-2}year^{-1}	3500
Efficiency PV cells (%)	15
Acreage of PV cells required to fulfil annual national electricity demand (ha)	60.000

solar radiation in winter is not enough to fulfil the electricity demand in this season and the surplus of generated electricity in summer has to be stored for use in winter. At present there are no energy storage facilities with capacities large enough to solve this seasonal problem. The introduction of PV-systems therefore requires large adaptations of the energy system. The system has to change from a stock driven system to a flow driven system and technological, economic and institutional changes are required. Comparison can be made with changes from energy carriers in the past. It took 100 years before coal had replaced wood as the major energy source, and 50 years before oil had replaced coal. (Figure 8.2). This implies that in the short term contribution of electricity generated by photovoltaic systems will be small.

4.2   Biomass

The prospects of electricity generated through combustion of biomass are much better. The big advantage of using biomass as an energy source is that variation in flux is smaller. In principle plant material is the result of the total incoming radiation over a growing season (and when trees are used radiation is even integrated over several years). The large variation in radiation levels, therefore does not affect the supply when biomass is used. Moreover this plant material can be stored, and supply can follow the demand. Further the woodchips can be combusted in presently existing electricity plants so that no major changes in the present energy system are required. Next to this the financial consequences of the use of biomass as feedstock for electricity plants are not insuperable. This makes electricity from biomass far easier to implement in the existing energy systems than electricity from PV-systems. Therefore electricity from biomass will be the most feasible option in the Dutch energy system in the short-term. However, the land claim introduced by the energy system will be very large. The policy goal that in 2020, 10% of the energy supply has to be fulfilled with renewable energy sources (Ministerie van Economische Zaken, 1995) implies that (using the 1999 values) 0.26 EJ has to be obtained from biomass. This means the use of 0.7 Mha for growing biomass crops; this is 20% of the total Dutch surface (and 35% of the land used for agricultural purposes).

5.   Discussion

The values used in the calculations above should be considered very rough estimates. In reality the system is far more complicated than described here. A major simplification is that Joules of electricity are assumed to be equal to Joules of heat. To give an example: PV-cells produce electricity, while a biomass plantation produces wood with a certain heating value. The conversion of this wood into electricity occurs with an efficiency of about 40%. This implies that it requires 2.5 times as much land area as calculated here to produce a certain amount of electricity from biomass. On the other hand this efficiency of 40% also accounts for fossil fuels that are converted into electricity. So when the goal is to replace 10% of fossil fuels by renewables, using the heat value for biomass is acceptable.

Renewable energy sources does not imply only biomass from plantations, for the Dutch system it is estimated that in 2020 about 50% of the renewable energy can be gained from other sources like wind energy and heat pumps (Ybema *et al.*, 1999). This implies that 0.13 EJ can be generated from other sources. Further biomass can also be obtained from so-called rest streams, which imply plant material from households, parks and forests. The quantities that can be gained are estimated to be 30 PJ= 0.03 EJ (Faaij *et al.*, 1997) for the Netherlands. Therefore not all biomass has to be obtained from short rotation forestry systems as is assumed in this chapter. On the other hand assumptions made with respect to yields that can be obtained are high (12 ton ha^{-1}) which is the potential production level. Presently obtained yields are much lower (7 ton ha^{-1}); when these yield levels are used in the calculation the area required will almost double.

So the uncertainty in values used here is a factor of two. But even if this large uncertainty is included in the evaluation, the area required for biomass in 2020 will be huge in comparison with the total land area in The Netherlands. It is not likely that this acreage will become available within The Netherlands in the coming decades, since all suitable land is in use for agricultural purposes. Therefore in studies of potentials for renewable energy sources in The Netherlands the import of large quantities of biomass is considered (Ybema *et al.*, 1999). The transportation of biomass requires energy (about 1 MJ ton^{-1} km^{-1} (by boat)). This implies that transportation distances should not be too large, since otherwise the energy gained in the biomass is lost in the transportation of it and there is no net CO$_2$ emission reduction. Latvia and Sweden are often mentioned as potential exporters. However, areas required for biomass do not change when biomass is produced somewhere else. In less densely populated countries, however, it will be easier to establish biomass plantations.

On a global scale land use differs strongly from the land use in The Netherlands: 50% of the surface is covered with forests and woodlands (in The Netherlands only 10%). On a global scale using a part of the present woodlands for growing biomass could be an option, in that case the interference with the food production does not exist. But the fact remains that vast amounts of land are required for the production of renewable energy. Which implies large changes in land use: from agricultural land or forest into short rotation forestry. Short rotation forestry systems act differently with respect to water use, fertiliser requirements, greenhouse gas emissions etc. compared to both forests and agricultural crops (Ledin, 1998). Implementation of large-scale short rotation forestry systems will have large effects on regional water availability and source-sink relations of greenhouse gasses and since good quality soils are required it also interferes with the food production.

Consequences of large scale biomass production for energy purposes on global change

Previous chapters of this book paid attention to land use and global change. In general present land-use changes are linked to food production. On a global scale these food production systems produce about 22 EJ of food. It is estimated that about 20% of the global NPP is affected by the food production systems (Wright, 1990). Further large changes in hydrological, carbon and nitrogen cycles are observed as a result of agricultural practices. The present global energy requirements for non-food energy (transport, heating etc) are a magnitude larger and mount up to 400 EJ (and are likely to rise in the future). When this 400 EJ has to be produced in the same system as is used for food (chemical conversion via photosynthesis into plant material), the impact of human consumption (food and energy) on global environment will increase. This increase will be smaller than the 20-fold rise that can be derived from the energy values mentioned. For food the energy content is of limited importance, nutrient values and tastiness also play an important role. These quality values do not account for biofuels so a more efficient conversion of solar radiation into plant material can be expected. However, since energy consumption is far larger than the food consumption it is very likely that impacts of biomass as major energy source will overrule the impacts of the present food production systems.

Not all biomass that can be used for energy originates from biomass plantations. In modern societies biomass reststreams are available that can be used for energy generation. The magnitude of the energy that can be generated from these reststreams is small in comparison with the quantities required (30 PJ for the Netherlands (Faaij et al., 1997), Haberl and Geissler (2000) estimated 78 PJ for Austria), however still large amounts of fossil fuels can be replaced with it. And since the use of these reststreams for energy generation has no effect on NPP (Haberl and Geissler (2000)) their use as an energy source is of importance. The impact of using PV systems as an energy source is likely to be smaller. In the first place since a smaller area is required but also since conversion follows another route (physical) with less interference with natural cycles. However, it should be realised that the storage problem with respect to PV systems is not yet solved. For an appropriate evaluation of the PV-systems the impact of the storage systems on the environment has to be included in the analysis.

It can be concluded that with respect to renewable energy supply, energy from biomass is the most appropriate solution for the short term (decade), since it fits best in the present energy system. The interaction of biomass production systems with natural cycles, however, makes the net effect of this renewable energy source on the global climate difficult to assess and great care should be taken in focusing on biomass as the most important future renewable energy source.

References

Broek, R., van den (2000) Sustainability of biomass electricity systems, Eburon, Delft.

CBS (2000) http://www.cbs.nl

EIA (2000) http://www.eia.doe.gov/emeu/iea/table11.html

Faaij, A.; Doorn,J.van; Curvers,T.; Waldheim,L.; Olsson,E.; Wijk, A.van; Daey Ouwens, C. (1997) Characteristics and availability of biomass waste and residues in the Netherlands for gasification, Biomass and Bioenergy,12 (4), 225–240.

Green, M.A. (1993) Crystalline- and polycrystalline-silicon solar cells, in: T.B. Johansson, H. Kelly, A.K.N. Reddy and R.H. Williams (eds), Renewable energy, sources for fuels and electricity: Washington, Island Press, 337–360.

Grubler, A., Nakicenovic, N. (1997) Decarbonizing the global energy system: Technological Forecasting and Social Change, 53, 97–110.

Haberl, H., Geissler, S. (2000) Cascade ustilization of biomass: strategies for a more efficient use of a scarce resource. Ecological Engineering 16, s111–s121.

Hall, D.O., RosilloCale, F., Williams, R H., Woods, J., (1993) Biomass for energy: supply prospects, , in: T.B. Johansson, H. Kelly, A.K.N. Reddy and R.H. Williams (eds), Renewable energy, sources for fuels and electricity: Washington, Island Press, 593–651.

KPMG (1999) Zonne-energie: een eeuwige belofte tot een concurrerend alternatief, (in opdracht van Greenpeace). KPMG bureau voor economische argumentatie, Hoofddorp.

Ledin,S. (1998) Environmental consequences when growing short rotation forests in Sweden: Biomass and Bioenergy, 15, 49–50.

Lysen E.H., Daey Ouwens, C., Onna, M. J. G., Blok, K., Okken, P A, Goudriaan, J. (1992) De haalbaarheid van de productie van biomassa voor de Nederlandse energie huishouding. Novem, Apeldoorn,

Ministerie van Economische Zaken (1995) Derde Energienota, Sdu Uitgevers, 's-Gravenhage.

Ministerie van Economische Zaken (1999) Energierapport 1999. Ministerie van Economische Zaken, Den Haag.

Nonhebel, S. (1997) Harvesting the sun's energy using agro-ecosystems. Quantitative Approaches in Systems Analysis no 13. AB-DLO, Wageningen.

Nonhebel, S. (2002) Energy yields in intensive and extensive biomass production systems, Biomass and Bioenergy (in press).

Velds, C.A. (1992) Zonnestraling in Nederland, Thieme, Baarn.

Wright, D.H. (1990) Human inpacts on energy flow through natural ecosystems, and implications for species endangerment. Ambio, 19, 189–194.

Ybema, J.R., Kroon, P., Lange, T.J., de, Ruijg, G.J. (1999). De bijdrage van duurzame energie in Nederland tot 2020. ECN-C-99-053. ECN, Petten.

Contributing Authors

Mirjam de Bruijn

African Studies Centre Leiden
P.O. Box 9555
2300 RB Leiden
The Netherlands

Wouter de Groot

Centre of Environmental Science
Leiden University
P.O. Box 9518
2300 RA Leiden
The Netherlands

Ali de Jong

Institute of Development Studies
Faculty of Geographical Sciences
University of Utrecht
P.O. Box 80115
3508 TC Utrecht
The Netherlands

Ton Dietz

Amsterdam Research Institute for Global Issues and Development Studies
University of Amsterdam
Nieuwe Prinsengracht 130
1018 VZ Amsterdam
The Netherlands

Han Dolman

Faculty of Earth and Life Sciences
Vrije Universiteit Amsterdam
De Boelelaan 1085
1081 HV Amsterdam
The Netherlands

Ronald Groeneveld

Environmental Economics and Natural Resources Group
Wageningen University and Research Centre
P.O. Box 16
6700 AA Wageningen
The Netherlands

Ronald Hutjes

Alterra
Wageningen University and Research Centre
P.O. Box 47
6700 AA Wageningen
The Netherlands

Jaap Huygen †

Alterra
Wageningen University and Research Centre
P.O. Box 47
6700 AA Wageningen
The Netherlands

Pavel Kabat

Alterra
Wageningen University and Research Centre
P.O. Box 47
6700 AA Wageningen
The Netherlands

Gideon Kruseman

Development Economics Group
Wageningen University and Research Centre
P.O. Box 16
6700 AA Wageningen
The Netherlands

Peter Kuikman

Alterra
Wageningen University and Research Centre
P.O. Box 47
6700 AA Wageningen
The Netherlands

Eddy Moors

Alterra
Wageningen University and Research Centre
P.O. Box 47
6700 AA Wageningen
The Netherlands

GertJan Nabuurs

Alterra
Wageningen University and Research Centre
P.O. Box 47
6700 AA Wageningen
The Netherlands

Sanderine Nonhebel

Center for Energy and Environmental Studies (IVEM)
University of Groningen
Nijenborgh 4
9747 AG Groningen
The Netherlands

Ruerd Ruben

Development Economics Group, Department of Social Sciences
Wageningen University and Research Centre
P.O. Box 8130
6700 EW Wageningen
The Netherlands

Han van Dijk

African Studies Centre Leiden
P.O. Box 9555
2300 RB Leiden
The Netherlands

Ekko van Ierland

Environmental Economics and Natural Resources Group
Wageningen University and Research Centre
P.O. Box 8130
6700 EW Wageningen
The Netherlands

Tom Veldkamp

Department of Environmental Sciences
Wageningen University and Research Centre
P.O. Box 37
6700 AA Wageningen
The Netherlands

Herman van Keulen

Plant Research International
Wageningen University and Research Centre
P.O. Box 16
6700 AA Wageningen
The Netherlands

Jan Verhagen

Plant Research International
Wageningen University and Research Centre
P.O. Box 16
6700 AA Wageningen
The Netherlands

Peter Verburg

Department of Environmental Sciences
Wageningen University and Research Centre
P.O. Box 37
6700 AA Wageningen
The Netherlands

Leo Vleeshouwers

Plant Research International
Wageningen University and Research Centre
P.O. Box 16
6700 AA Wageningen
The Netherlands

Fred Zaal

Amsterdam Research Institute for Global Issues and Development Studies
University of Amsterdam
Nieuwe Prinsengracht 130
1018 VZ Amsterdam
The Netherlands

Index